Advances in Anatomy
Embryology and Cell Biology

Vol. 165

Editors

F. Beck, Melbourne B. Christ, Freiburg
W. Kriz, Heidelberg W. Kummer, Gießen
E. Marani, Leiden R. Putz, München
Y. Sano, Kyoto T. H. Schiebler, Würzburg
G. C. Schoenwolf, Salt Lake City
K. Zilles, Düsseldorf

Springer

Berlin
Heidelberg
New York
Barcelona
Hong Kong
London
Milan
Paris
Tokyo

N. Ulfig

Calcium-Binding Proteins in the Human Developing Brain

With 44 Figures and 2 Tables

 Springer

N. Ulfig

AG Neuroembryologie,
Institut für Anatomie der Universität Rostock,
Gertrudenstr. 9, 18057 Rostock,
Germany
(email: norbert.ulfig@med.uni-rostock.de)

ISBN-13:978-3-540-43463-4
e-ISBN-13:978-3-642-59425-0
DOI: 10.1007/978-3-642-59425-0

Springer-Verlag Berlin Heidelberg New York
Library of Congress-Cataloging-in-Publication-Data
Calcium-binding proteins in the human developing brain / N. Ulfig. p. cm.
(Advances in anatomy, embryology, and cell biology; Vol. 165)
  Includes bibliographical references and index.
ISBN-13:978-3-540-43463-4
1. Calcium-binding proteins. 2. Moleculaar neurobiology. I. Title. II. Series.

This work is subject to copyright. All rights are reserved, whether the whole or part of the material is concerned, specifically the rights of translation, reprinting, reuse of illustrations, recitation, broadcasting, reproduction on microfilms or in any other way, and storage in data banks. Duplication of this publication or parts thereof is permitted only under the provisions of the German Copyright Law of September 9, 1965, in its current version, and permission for use must always be obtained from Springer-Verlag. Violations are liable for prosecution under the German Copyright Law.

Springer-Verlag a member of BertelsmannSpringer
Science + Business Media GmbH

http://www.springer.de
© Springer-Verlag Berlin Heidelberg 2002

The use of general descriptive names, registered names, trademarks, etc. in this publication does not imply, even in the absence of a specific statement, that such names are exempt from the relevant protective laws and regulations and therefore free for general use.

Product liability: The publishers cannot guarantee the accuracy of any information about dosage and application contained in this book. In every individual case the user must check such information by consulting the relevant literature.

Production: PRO EDIT GmbH, 69126 Heidelberg, Germany
Printed on acid-free paper   SPIN: 10867420   27/3130Re – 5 4 3 2 1 0

Contents

Acknowledgments

The author wishes to thank S. Cleven, J. Müller, F. Neudörfer and G. Ritschel for typing the manuscript and for preparation of the figures.

1 Introduction

1.1
Calcium-Binding Proteins

In the mature brain, calcium ions play pivotal roles in transmembrane and intracellular transmission of signals. Thus, calcium is involved in numerous neuronal functions including neurotransmitter release, enzyme regulation, modulation of neuronal excitability, gene expression, microtubular transport, or synaptic plasticity (Miller 1991). Many of these calcium-dependent processes are mediated or modulated by a number of cytosolic calcium-binding proteins (CaBPs). All nerve cells contain the calcium-binding protein calmodulin. Other CaBPs are restricted to certain nerve cell types, i.e., parvalbumin (PV), calbindin (CB) and calretinin (CR) (Baimbridge et al. 1992).

These proteins are members of the EF-hand family of CaBPs. The EF-hand, representing an amino acid sequence with a characteristic three-dimensional structure, is the high-affinity calcium-binding site.

The small water soluble PV was first isolated from skeletal muscles (Heizmann 1984) and has three EF-hand domains, two of which are functional. In mammalian neurons only one isoform (i.e., the α-isoform) exists (Celio and Heizmann 1981). The gene for PV, having a molecular weight of about 12 kDa, is located on chromosome *22* in humans (Berchtold 1989).

The CaBP CB, which was first isolated from chicken intestine mucosa, exists in several forms, i.e., 6 kDa, 8–11 kDa, 9 kDa, and 28 kDa. The 28-kDa CB occurs in distinct neuronal populations and has six putative EF-hand domains, four of which are thought to be functional (Heizmann 1984). In the intestine and the kidney of different species, the synthesis of CB is induced by vitamin D (Christakos et al. 1979). Such an induction could not be shown in the rodent brain (De Viragh et al. 1989). The gene for CB is located on chromosome 8 (Parmentier 1980).

The CaBP CR was detected when an antibody directed against CB was found to cross-react with a slightly larger molecule in homogenates of rat brain (Pochet et al. 1985). Subsequently, CR has been identified as a CaBP which structurally is closely related to CB. CB and CR share about 60% of their amino-acid sequence (Pasteels et al. 1987). Genes coding for CB and CR are located on different chromosomes (i.e., CR on chromosome *16*, CB on chromosome *8*) in humans. Both genes are situated in close vicinity to carbonic anhydrase isoenzyme genes. Therefore, a common duplication of CB/CR and carbonic anhydrase isoenzyme genes appears likely (Parmentier et al. 1991).

1.2
CaBPs in the Mature
and Developing Central Nervous System

A wealth of data is found in the literature demonstrating that the three CaBPs, PV, CB, and CR, occur in distinct subsets of neurons in the central nervous system (CNS) (for review, see Baimbridge et al. 1992). Therefore antibodies raised against PV, CB, and CR have been used extensively as markers of neuronal classes or subclasses. Only partial overlap in the distribution patterns of the CaBPs is observed.

Because of their high solubility, the three CaBPs are present in the entire soma and in the processes of neurons. Thus, immunostaining with antisera against PV, CB, and CR yields a Golgi-like appearance of nerve cells, allowing neuroanatomists to carry out classification of nerve cells (Heizmann and Brown 1992).

The CaBPs CR and CB are expressed during early development. Mostly, CR appears most precociously. PV, however, occurs distinctly later during ontogenesis and has, therefore, been associated with mature neuronal activity, whereas CB and CR may control developmental events (Hendrickson et al. 1991; Solbach and Celio 1991; Ellis et al. 1991; Enderlin et al. 1987).

Expression of the three CaBPs appears in given brain areas at definite times of development. The distribution patterns of the CaBPs often differ from those found in the mature brain. Such transient expression patterns imply that an increase, decrease, or resolution of CaBP-immunoreactive (ir) structure takes place during a later period in development (Andressen et al. 1993). The transient expression of CaBPs may be linked to key developmental events such as cell movement (i.e., migration) or process outgrowth.

Calcium is involved in various aspects of neuronal differentiation such as gene transcription, enzyme activation, synaptogenesis, pattern formation, transmitter phenotype, axonal elongation (Walicke and Patterson 1981; Cline and Tsien 1991; Dash et al. 1991; Ghosh et al. 1994). On account of their capacity to regulate intracellular calcium concentrations CaBPs are important proteins during neuronal differentiation.

1.3
Overview:
Features of the Developing Brain

The first major step in brain development is the generation of large numbers of neuroblasts within the ventricular zone. At first this ventricular zone occupies the entire thickness of the CNS wall. Another key event during CNS development is the migration of postmitotic neurons from the proliferative zone to their final targets. Migration is guided by long glial fibers which are radially oriented. These radial glial cells span their processes along the full width of the cerebral wall and thus provide scaffolding for migrating neurons (Rakic 1988, 1995). The radial glial cells form close contacts, i.e., surface-mediated interactions, with the migrating neurons of a bipolar shape. This predominantly radial migration suggests that region-specific differences in the cerebral cortex may be specified early in development. A site within the proliferative zone gives rise to neurons, all of which migrate along the same radial glial fascicles and from an ontogenetic column within the cortical plate. Thus, each column consists of cells that originated from the same proliferative unit (Rakic 1974; Rakic 1995).

Recently, evidence has been provided that a significant number of cortical neurons reach their final destination by tangential (nonradial) migration. Thus, neurons produced in a definite domain of the proliferative zone are widely dispersed throughout the cerebral cortex via extensive tangential migration. It is most likely that the two different major types of cortical neurons, i.e., pyramidal projection neurons and gamma-aminobutyric acid (GABA)-ergic interneurons, take different migratory routes to their final targets: Pyramidal neurons migrate radially, whereas interneurons undergo extensive tangential migration from the ganglionic eminence to various areas of the cerebral cortex (Parnavelas 2000).

Neurons of the diencephalic nuclei are produced in the ventricular zone lining the third ventricle. It is well known that the neurons of lateral diencephalic nuclei originate first from the ventricular zone and that the subsequently produced neurons settle more medially ("outside-in" gradient) (Kostovic 1990a).

The first processes which arise from the developing neurons are all alike and grow slowly. Eventually one process accelerates its growth rate and becomes the axon. Outgrowing axons are guided toward their appropriate target by a great number of molecular cues such as components of the extracellular matrix or cell adhesion molecules. Growth of axons can also be directed by diffusible chemoattractants. This mechanism appears to be specific for certain classes of axons (Brown et al. 2001; Oudega et al. 1993). After a growth cone has met, a suitable target synapse formation arises.

The structural organization of the fetal brain differs substantially from that of the mature brain and it is continuously changing (Fig. 1). The fetal brain displays conspicuous zones that are unique to the developing brain and have no direct counterparts in the adult. During fetal development, the cerebral pallium shows a typical pattern of lamination consisting of the following zones (from ventricle to pallium): ventricular zone, subventricular zone (second proliferative zone), intermediate zone (future white matter), subplate, cortical plate, and marginal zone (Kostovic 1990b; Ulfig 2000b). The fetal layers reflect the transient arrangements of fibers, synapses, and nerve cells. The content of each zone is permanently changing. The entire fetal development represents a period of great dynamics (Chan et al. 2002).

A pronounced feature of the human developing brain during the second half of gestation is the establishment of transitory neuronal circuitries. The latter are essential for building up appropriate mature projections. The fetal transitory circuitries are associated with transient structures which are outstanding components of the human developing brain and often act as intermediate targets for outgrowing axons (Ulfig et al. 2000c; Ulfig and Chan 2002a).

1.4
Scope of this Review

The great majority of studies on CaBPs in the mature and developing brain have been performed on animals. The present volume of *Advances in Anatomy and Embryology* focuses on the distribution patterns of CaBPs in the developing human brain. Special

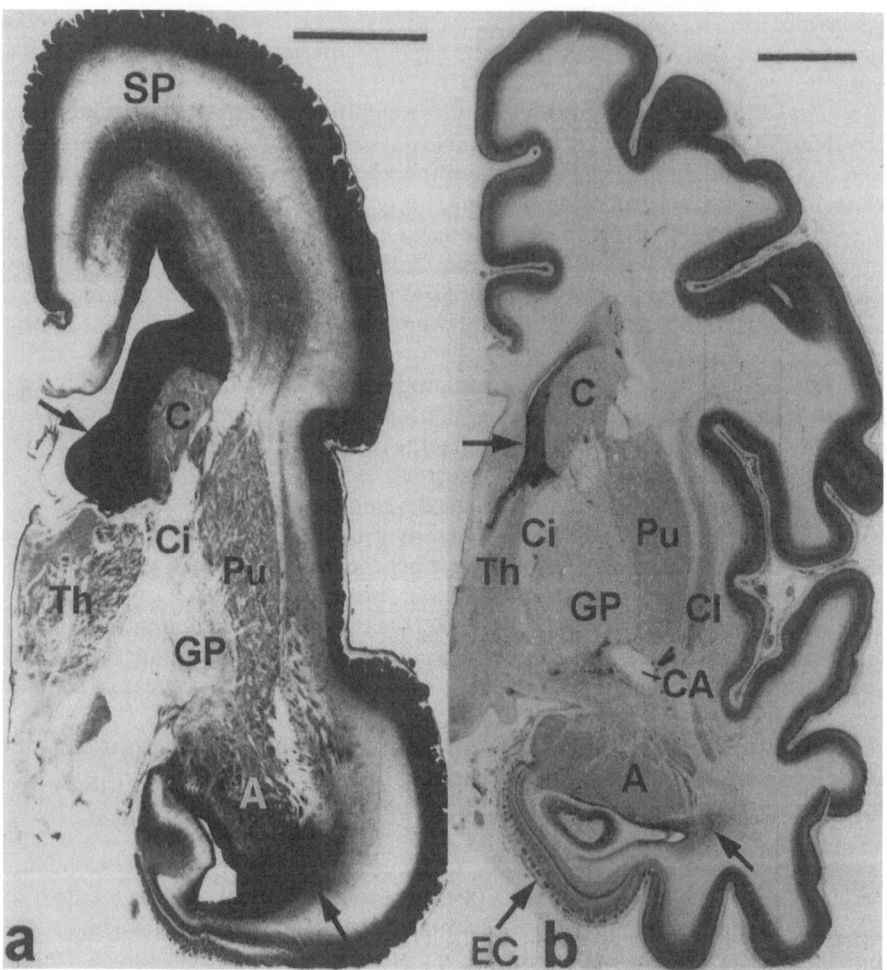

Fig. 1a, b. Frontal sections of the human fetal brain, Nissl-stained: **a** 5th gestational month, **b** 8th gestational month. *Arrows* mark the ganglionic eminence. *A*, amygdala; *C*, caudate nucleus; *CA*, anterior commissure; *Ci*, internal capsule; *Cl*, claustrum; *EC*, entorhinal cortex; *GP*, globus pallidus; *Pu*, putamen; *SP*, subplate; *Th*, thalamus. Scale bars: a, 3 mm, b, 5 mm. (From Ulfig et al. 2000c, with permission)

attention is paid to the expression of CaBPs in transient structures of the human fetal brain, to developmental changes in the expression patterns, and to subtle changes in pathologically altered specimens. Thus, new insights into the transitory functional organization unique to the human fetal brain are summarized in this monograph and are discussed with regard to neuronal circuitries in the developing brain and their relevance in fetal and perinatal brain injury.

1.5
Materials and Methods

A total of 90 human brains have so far been used for studies on the distribution patterns of CaBPs. For details concerning the brains, see Table 1. The brains were obtained from legal elective and spontaneous abortions according to German laws with the prior consent of an ethical committee. The postmortem delays ranged from 1 to around 32 h. From the whole brain, the brain stem was detached at the level of the inferior mesencephalon and the hemispheres were cut in the mediosagittal plane. The brains were fixed by immersion in 3.7% paraformaldehyde (pH 7.4) for around 2 days on a rotator. The hemispheres were cut into frontal blocks and cryoprotected in 0.05 M TRIS-buffered saline (TBS, pH 7.4) containing 30% sucrose, until the descent of the specimens to the bottom of the solution indicated sufficient cryoprotection. Then the blocks were frozen and cut into 120-µm-thick sections on a cryostat. Free-floating sections were always rinsed three times in TBS after cutting and after each step of the immunohistochemical staining procedure (if not stated otherwise, see double-labelings). All incubation steps were performed on a rotator.

Reduction of nonspecific staining was carried out (1) in TBS containing 10% methanol and 7% hydrogen peroxide (30 min) and (2) in TBS containing 1.5% lysine, 10% bovine serum albumin (BSA), and triton X-100 (2 h). Then the sections were incubated in the primary antibody at 4°C for 2 days. Primary antibodies and their dilutions are listed in Table 2. Thereafter, sections were transferred to the biotinylated secondary antibody (anti-mouse IgG or anti-rabbit IgG, diluted 1:200 in TBS with 2% BSA) for 90 min. Then the avidin-biotin-peroxidase complex (ABC, diluted 1:25 in TBS containing 2% BSA) was allowed to react for 2 h. Subsequently the immunocomplex was visualized using diaminobenzidine (0.07% in TBS) and 0.003% hydrogen peroxide. The immunostained sections were mounted on gelatin-coated slides, dehy-

Table 1. Brains used for the studies reviewed in this monograph

Age (weeks of gestation)	CNS diagnosis	Number (n)
16–20	Normal	11
21–25	Normal	18
26–30	Normal	20
31–40	Normal	16
Postnatal, 1–3 months	Normal	4
Adult, 78/80 years	Normal	2
21–24	Hydrocephalus	7
32–36	Hydrocephalus	4
18–25	Trisomy 22	2
18 -25	Trisomy 21	4
23	Trisomy 18	2

Table 2. Antibodies used for the studies reviewed in this monograph

Antibodies	Host	Dilution	Provider
AKAP79	Mouse	1:500	Transduction Laboratories, Lexington, KY, USA
Calretinin	Rabbit	1:5000	SWANT, Belinzona, Switzerland
Calbindin1	Rabbit	1:2000	SWANT, Belinzona, Switzerland
	Mouse	1:2000	SWANT, Belinzona, Switzerland
GAP 43	Mouse	1:1000	Sigma, St. Louis, MO, USA
SMI 35	Mouse	1:1000	Sternberger Monoclonals, Baltimore, MD, USA
SNAP 25 (SMI 81)	Mouse	1:1000	Sternberger Monoclonals, Baltimore, MD, USA
Synaptophysin1	Mouse	1:200	Sigma, St. Louis, MO, USA
	Mouse	1:100	Boehringer, Mannheim, Germany
Parvalbumin1	Rabbit	1:5000	SWANT, Belinzona, Switzerland
	Mouse	1:5000	SWANT, Belinzona, Switzerland

drated in a graded series of alcohols, cleared in xylene, and coverslipped (for further details, see Ulfig et al. 1998a).

For the double-labeling procedure, preincubation steps were identical to those described for single-labelings (see preceding paragraph). Incubation was carried out with two antibodies of different species (Table 2). Then peroxidase- (diluted 1:200) and alkaline phosphatase-(diluted 1:100) labeled secondary antibodies were applied for 2 h. First, the peroxidase was visualized using 0.07% DAB and 0.003% hydrogen peroxide. After three rinses in phosphate buffer (pH 8.6) containing 0.025% levamisole, the alkaline phosphatase was visualized with 5-bromo-4-chloroindolyl-phosphate/nitroblue tetrazolium (BCIP/NBT) diluted 1:50 in TBS. After rinsing in distilled water, sections were mounted on gelatin-coated slides and coverslipped with glycerol gelatin (for further details, see Ulfig et al. 1998b).

Method specificity was proved by the absence of specific immunolabeling when omitting the primary antibody.

Every tenth cryostat section was not immunostained but stained with cresyl-violet for demonstrating Nissl substance. These Nissl-stained sections were used for orientation; thus, borders of nuclei, layers, or areas were determined. Frequently, camera lucida drawings of these borders were made, then the immunoreactive (ir) structures were plotted on these drawings from the serially adjacent immunopreparations.

Regularly, photographic documentation was made at low and high magnification. Furthermore, camera lucida drawings of neuronal types were made at high magnification.

The section thickness used for the studies reviewed here is considerably higher than that customarily used. Thus, the fine details of neurons and their processes as well as the orientation of the neurons are visible at different levels of focus. Moreover, in these relatively thick sections nerve cells superimpose on each other; thus, borderlines of nuclei or layers stand out more clearly.

In general, incubation of free-floating sections was preferred to immunostaining procedures performed on mounted sections. The latter display more background staining and less intense specific immunolabeling than free-floating sections.

2 Cerebral Cortex: Subplate

2.1
Description

The first cortical neurons separate from the proliferative zone (i.e., ventricular zone) and build up a single layer on the nonventricular side called primitive plexiforme layer or preplate. Then a third zone becomes visible between the ventricular zone and the preplate, i.e., the intermediate zone. Neuronal precursors destined to form the cortical plate migrate through the intermediate zone and the lower half of the pre-plate. They settle in the middle of the preplate and thus divide the latter into two layers: the outer layer becomes the marginal zone (layer I of the mature cortex) and the deeper becomes the subplate (Marin-Padilla 1998a; Kostovic and Rakic 1990).

Around 13 weeks of gestation, the deep part of the cortical plate becomes loose; this loosening indicates that cells of the cortical plate become incorporated into the subplate. Neurons derived from the cortical plate are likely to form the upper part of the subplate (Kostovic and Rakic 1990). Moreover, neuronal precursors arising from the ganglionic eminence settle within the subplate.

An extensive increase in thickness of the subplate is observed until 22 weeks of gestation: the amount of fibers dramatically increases and all the neuronal types characteristic of the subplate occur (Ulfig et al. 2000c; Kordower and Rakic 1990).

In Golgi preparations, four neuronal types can be distinguished within the sub-plate, according to Mrzljak et al. (1988) and Kostovic and Rakic (1990):
1. *Polymorphic neurons*. Their cell bodies display various shapes and their dendrites originating from random sites of the somata show bifurcation patterns.
2. *Fusiform neurons*. From the spindle-shaped soma of these neurons, two long and thick dendrites emerge. The axons usually originate from one of the two main dendrites. These nerve cells can be oriented in all directions.
3. *Pyramidal neurons*. Around 80% of the pyramidal neurons reveal an entirely inverted cell body so that the axons originate from the inverted base or from the basal dendrites and run towards the cortical plate. In comparison to pyramidal neurons in the cortical plate, the subplate pyramidal cells reveal a smaller cell body and less dendritic branches.
4. *Multipolar neurons*. They are characterized by spherical cell bodies and stellate dendritic trees, the latter of which emerge from three to five stem dendrites.

The four neuronal types can be clearly distinguished from young neurons migrating through the subplate. Migrating neurons show a smaller soma size, a bipolar shape, and a perpendicular orientation. On the whole, the nerve cell types of the subplate reveal a relatively extensive dendritic surface, which facilitates their involvement in various fetal circuitries. In general, the nerve cells in the subplate are characterized by a precocious morphological and neurochemical differentiation.

The neurons of the subplate express a number of neuroactive substances such as GABA, neuropeptide Y, microtubule-associated protein (MAP) 2, somatostatin, substance P and cholecystokinin (Meinecke and Rakic 1992; Mehra and Hendrickson 1993; Kostovic et al. 1991; Chun et al. 1987). Neurons immunolabeled by various antibodies are not evenly distributed within the subplate zone. Furthermore, the time frames during which the above-mentioned substances are expressed differ to some extent.

2.2
Expression of CaBPs in the Subplate

The CaBPs PV, CB, and CR are found within the subplate. The expression of PV is only moderate; PV-ir neurons are restricted to the upper third of the subplate and occur rather late in development. At 30 weeks of gestation a few medium-sized and small PV-ir nerve cells displaying various shapes are observed.

CR and CB are abundantly expressed within the subplate (Fonseca et al. 1995; Liu and Graybiel 1992b; Enderlin et al.1987). Thus, antibodies directed against these two CaBPs have been suggested as adequate markers of the subplate.

Using anti-CR, the subplate is intensely immunolabeled in the 5th and 6th gestational month. Thus, the borders towards the cortical plate and the intermediate zone can be clearly defined (Fig. 2). During later developmental stages, the packing-density of CR-ir cells is distinctly reduced. From 30 weeks of gestation onwards, only a small number of CR-ir nerve cells are encountered in the upper third of the subplate.

CB expression is observed in the subplate from around 25 weeks of gestation onwards. CB-ir nerve cells are particularly concentrated in the upper third of the subplate (Fig. 3). These neurons are a heterogeneous neuronal population with regard to their morphological appearance. A high number of CB-ir neurons belong to the class of polymorphic neurons which display various shapes and do not exhibit a preferential orientation (Fig. 4).

Moreover, fusiform neurons, which are oriented in all directions, and multipolar neurons are present. The multipolar neurons show a stellate dendritic tree. The intensity of immunolabeling varies to a considerable extent. Some of the CB-ir neurons appear triangular and resemble pyramidal cells.

2.3
Functional Roles of the Subplate

During fetal development, the width of the subplate increases considerably in size, most probably due to an ingrowth of fibers. The subplate contains axons of various cortical afferent systems. These axons establish synaptic contacts with subplate neurons and reside in this zone for a prolonged period of time prior to entering the cortical plate. Therefore, the subplate is commonly described as a waiting compartment. It mediates the sequential and orderly formation of connections between vari-

Fig. 2. Frontal section of the occipital lobe, CR immunostained, 25th gestational week. Borders between the cortical plate (*CP*) and the subplate (*SP*) as well as between the subplate (*SP*) and intermediate zone (*IZ*) are marked by *arrows*. Scale bar: 5 µm. (From Ulfig et al. 2000c, with permission)

CP

SP

IZ

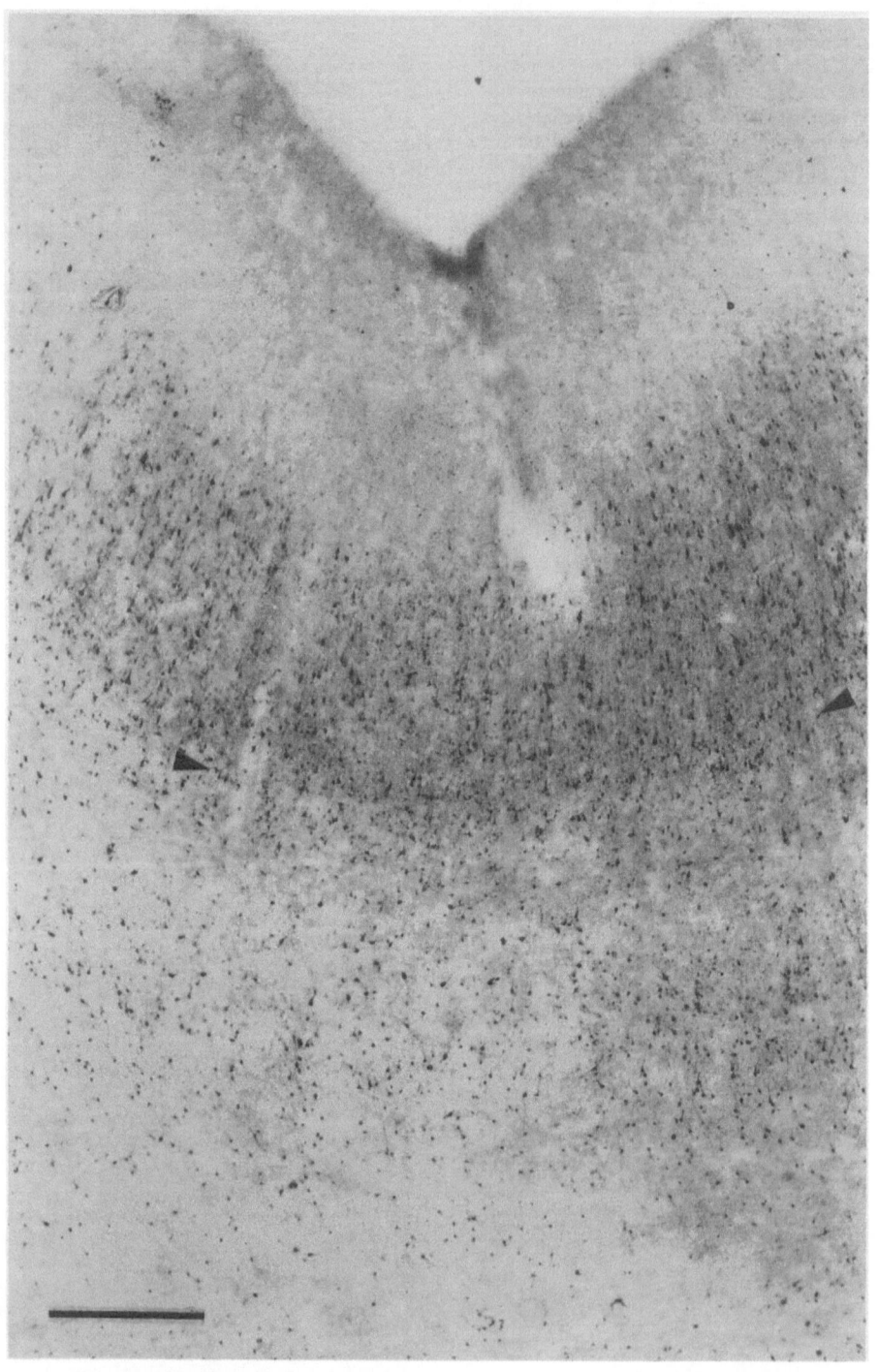

Fig. 3. CB immunopreparation of the cortical plate and subplate. *Arrowheads* mark the border of the two zones. Scale bar, 400 μm

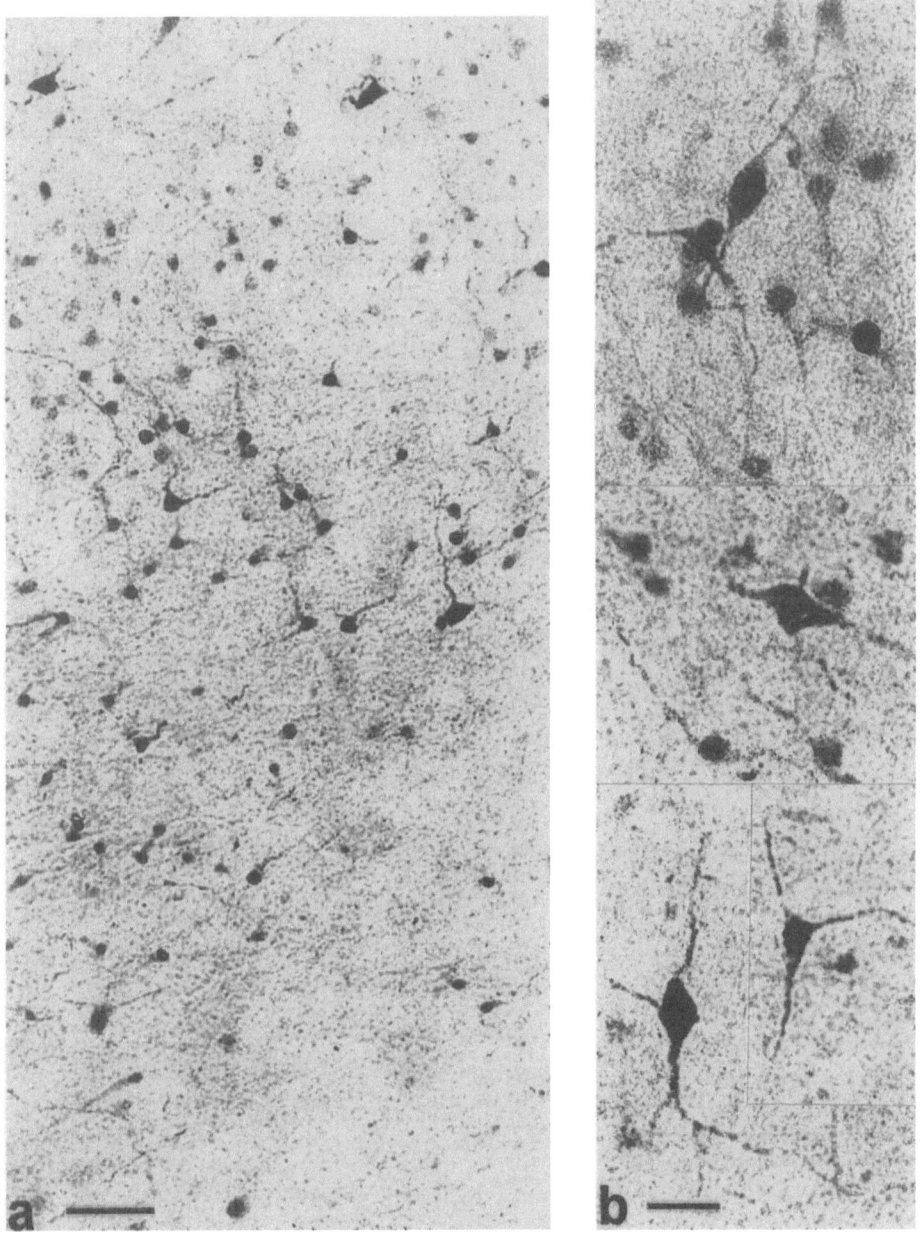

Fig. 4a, b. CB-ir neurons of the subplate (*upper third*) at survey magnification (**a**) and at high magnification (**b**), 28 weeks of gestation. Scale bars: **a**, 50 μm; **b**, 20 μm

ous subcortical nuclei and the appropriate cortical target regions (Ulfig et al. 2000c). First, monoaminergic fibers coming from brain stem nuclei arrive within the subplate (Zecevic 1993; Verney et al. 1995). The second fiber system to arrive in the subplate stems from the cholinergic magnocellular nuclei of the basal forebrain (Kostovic 1986).

Subsequently, thalamic fibers arrive within the subplate. After selective elimination of subplate neurons, these thalamic fibers fail to enter the appropriate cortical areas (Allendoerfer and Shatz 1994). The last fibers to arrive in the subplate are those stemming from ipsi- and contralateral cortical areas. These fibers remain the longest and form a large contingent of fibers in the subplate (Kostovic and Rakic 1990).

The subplate contains projection neurons which send their axons towards subcortical targets (McConnell et al. 1989). These pioneer axons can provide a scaffold which may guide thalamocortical fibers and corticofugal axons to their subcortical targets. In addition to these descending axons, the subplate sends projections to the cortical plate (Friauf et al. 1990). These projections could be required for the refinement of thalamic connections within lamina IV (Ghosh and Shatz 1992). Moreover, subplate neurons have been shown to project to the contralateral hemisphere (Chun et al. 1987; Chun and Shatz 1989).

Additionally, the subplate neurons build up local circuits. These local circuit neurons, which are GABAergic, may influence transient connections within the subplate (Meinecke and Rakic 1992).

Furthermore, experimental evidence has been provided that the subplate is involved in the formation of cerebral gyri. This relationship between the subplate and cerebral convolutions is substantiated by an observation in the human brain. Tertiary gyri of the frontal lobe only develop postnatally; in these locations the subplate persists for a longer period postnatally (Kostovic et al. 1989).

2.4
Resolution of the Subplate

The subplate diminishes and eventually disappears at a distinctly slow rate. The exact time point of the resolution is difficult to determine as some subplate neurons persist and are found as interstitial neurons within the adult white matter. The resolution of the subplate is, moreover, marked by the departure of terminals (Kostovic and Rakic 1990).

Dilution in the increasing volume of the white matter is likely to contribute to the decrease in neuronal density of the subplate (Wood et al. 1992). Moreover, evidence has been provided that nerve cell death is a major mechanism underlying the resolution of the subplate (Allendoerfer and Shatz 1994; Kordower and Mufson 1992).

2.5
Effect of Fetal Hydrocephalus
on the Distribution Pattern of CaBPs in the Subplate

Hydrocephalus can produce dramatic changes in the thickness of the cerebral mantle. Ventricular expansion in the fetal brain can result in an increase in cortical circumference due to unfused sutures (Sainte-Rose 1993). Thus, the combination of compression and stretching is likely to cause severe damage in the brain at a period when important developmental events take place.

In hydrocephalic brains (age, 32–34 weeks of gestation), distinct alterations in the distribution pattern of CB and PV can be observed in the subplate of the occipital lobe in comparison to controls (Ulfig et al. 2001b). The occipital lobes have been divided into two blocks. The cortical mantle of the anterior occipital block is severely thinned, whereas that of the posterior block is extremely thinned.

In CB- and PV-immunosections of severely altered hydrocephalic tissue, ir nerve cells are observed at the same location within the subplate as in controls (Fig. 5). Also, the number of ir nerve cells does not differ in hydrocephalus as compared to controls. However, at higher magnification, differences in nerve cell morphology are apparent: the CB and PV nerve cells are distinctly smaller in hydrocephalus and dendritic trees are not immunolabeled in contrast to controls.

In CB and PV immunopreparations of extremely altered hydrocephalic tissues, no ir structures are found in the subplate.

It may be assumed that the changes found in hydrocephalic brains may interfere with normal neuronal circuitry in the fetal brain. Thus, an impairment of transient circuitries may disturb the development of mature cortical connections.

As the subplate is involved in the formation of cerebral gyri, it is tempting to assume that the disturbance of the subplate function may be linked to the appearance of small irregularly arranged convolutions, which are frequently observed in fetal hydrocephalic brains. So far, the development of this redundant gyration (polymicrogyria) has not been consistently explained (Friede 1989).

The reduction or loss of CB and PV may cause a significant disturbance of the intracellular calcium homeostasis. Calcium is involved in various developmental processes such as cell movement, morphological and functional differentiation (Spitzer 1994), and axonal growth (al-Mohanna et al. 1992; Mattson and Kater 1987). Moreover, calcium activates various enzymes, promotes the disassembly of microtubules, participates in the regulation of synthesis and release of neurotransmitters, and regulates fast axonal transport and membrane excitability (Cheung 1980; Gupta and Dudani 1989; Hammerschlag et al. 1975; Hinrichsen et al. 1986). Thus, a decrease in expression of CaBPs which participate in calcium-mediated signaling mechanisms is most likely to reflect a decline in neuronal function.

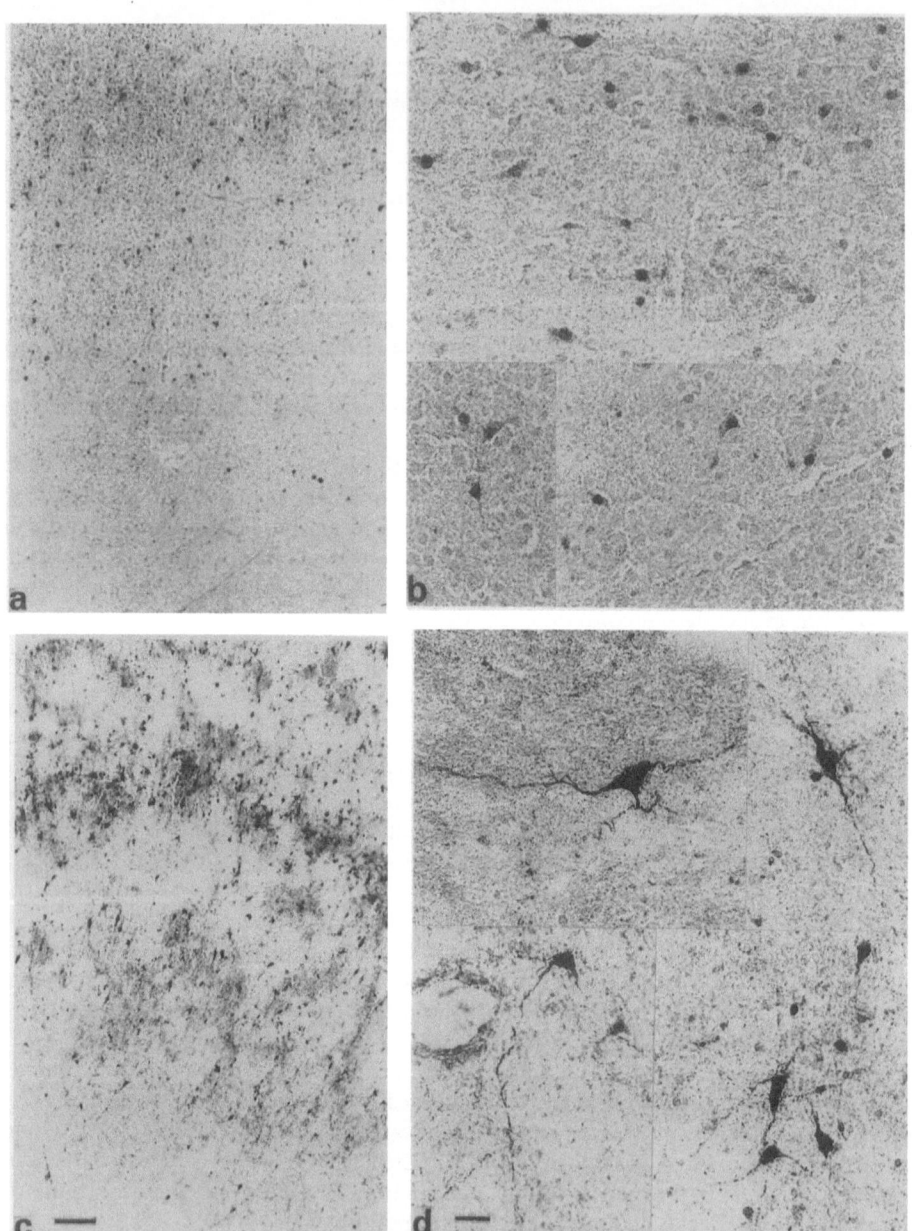

Fig. 5a–d. CB immunopreparations of the subplate in the occipital lobe, 32 weeks of gestation. *a, b* Severely altered hydrocephalic tissue. **c, d** Control case. Scale bars: a and c, 100 μm; b and d, 10 μm. (From Ulfig et al. 2001b, Karger Basel)

2.6
Significance of the Subplate in Preterm Infants

The possible importance of the subplate in brain injury of the premature infant has been postulated by Volpe (1996). The peak development of the subplate closely matches the time frame when ischemic lesions occur in preterm infants. Thus, important functions of the subplate could be impaired. This assumption would enable one to find a morphological substrate of cognitive deficits which occur in 25%–50% of small premature infants.

Antisera against CaBPs provide a useful tool for detailed investigation of neuronal alterations in pathologically altered specimens of preterm infants.

3 Cerebral Cortex: Cortical Plate

3.1
Description

The layers of the cerebral cortex are generated in an orderly sequence. The precursor cells of the deepest layers leave the proliferative zone to migrate first to the cortical plate. Thus, the laminar position of the migrating cell depends on when it was generated. Neurons that are generated late have to pass neurons generated earlier to settle more superficially. This spatiotemporal correlation between the time of generation and the laminar position of a neuron is described as the "inside-out gradient" (Brown et al. 1991). This concept of an "inside-out gradient" in neurogenesis was established at a time when the assumption that all cortical neurons migrate radially to their final positions was generally accepted. Recently, various studies have shown that the majority of interneurons have their origin in the ganglionic eminence. Thus, these precursor cells have to migrate tangentially from the ganglionic eminence to their final position in the cerebral cortex (Lumsden and Gulisano 1997; Parnavelas 2000).

Two principal neuronal classes are found in the cerebral cortex, pyramidal projection neurons, and nonpyramidal interneurons. The latter all contain the inhibitory transmitter γ-aminobutyric acid (GABA).

3.2
CaBPs in the Cortical Plate

Within the class of inhibitory interneurons, various neuronal subsets can be defined according to their differential expression of CR, CB, and PV.

Between 21 and 25 weeks of gestation, a high number of CR-ir neurons is seen in the cortical plate. They are evenly distributed and no differences in number are observed when comparing deeper to superficial areas of the cortical plate. The small CR-ir nerve cells show a bipolar, sometimes a triangular cell body. Two or three processes are seen to emerge from the soma. The majority of ir nerve cells appear undifferentiated. Some triangular nerve cells exhibit a medium-sized soma and processes that bifurcate; thus, these neurons appear more mature.

Between 32 and 36 weeks of gestation, a large number of evenly distributed CR-ir nerve cells are present in the cortical plate. The medium-sized neurons display a bipolar, triangular or multipolar shape, and their processes can be traced up to 50 µm.

This distribution pattern of CR-ir neurons closely corresponds to findings described in the rat cerebral cortex (Fonseca et al. 1995).

Between 21 and 25 weeks of gestation, no PV-ir structures are encountered in the cortical plate. Between 32 and 36 weeks of gestation, a high number of PV-ir nerve cells are observed in deep layers of the cortical plate. In this location they appear densely packed. The majority of PV-ir neurons exhibit a multipolar shape. Their processes can regularly be traced up to 100 μm.

This distribution pattern of PV-ir neurons closely matches that described for the cat cerebral cortex by Stichel et al. (1987). These authors also found that the PV-ir bulb is concentrated in lower layers and that PV expression appears rather late in development.

Like PV expression, CB expression is not observed in the cortical plate between 21 and 25 weeks of gestation.

Between 32 and 36 weeks of gestation, CB-ir neurons are observed within the entire cortical plate; their packing density, however, is distinctly higher in the deeper layers (Fig. 3). Nerve cells in the superficial layers appear less differentiated than those in the deeper layer. Superficial CB-ir cells are small and dendrites can only be traced over short distances. Deeper CB-ir neurons exhibit a larger soma size and a more elaborated dendritic tree (Fig. 6). A similar distribution pattern of CB-ir neurons has also been observed in the cat cerebral cortex at later developmental stages (Stichel et al. 1987).

3.3
Effect of Fetal Hydrocephalus on the Distribution Pattern of CaBPs in the Cortical Plate

When comparing the number and distribution patterns of CR-ir neurons in the cortical plate of controls with those of hydrocephalic brains, no differences are evident (Ulfig et al. 2001b). However, in hydrocephalic brains, neuronal somata are distinctly smaller and only stems of dendritic trees are immunostained.

In PV immunopreparations of severely altered specimens (for definition, see Sect. 2.5), the number of PV-ir neurons in the cortical plate is moderately decreased. Moreover, neuronal cell bodies appear smaller, their shape is in general bipolar or triangular, and only the stems of processes are immunolabeled. Thus, a Golgi-like appearance of PV-ir neurons, as seen in controls, is not visible in severely altered specimens of hydrocephalic brains.

In PV immunopreparations of extremely altered tissue, no ir structure can be detected.

In CB immunosections of severely altered specimens, the number of ir neurons is moderately decreased. Neurons appear smaller and their processes are not or only weakly immunolabeled. Bifurcations of dendrites, regularly seen in controls, are not observed in severely altered tissue.

In CB immunopreparations of extremely altered specimens, no ir structures are found.

The gradient in loss of PV- and CR-ir neurons observed when comparing severely and extremely altered tissue (Fig. 7) is in line with the general observation in hydrocephalic brains that the magnitude of ventricular enlargement influences the severity of neuronal damage (Del Bigio 1993). In experimental infantile hydrocephalus, reac-

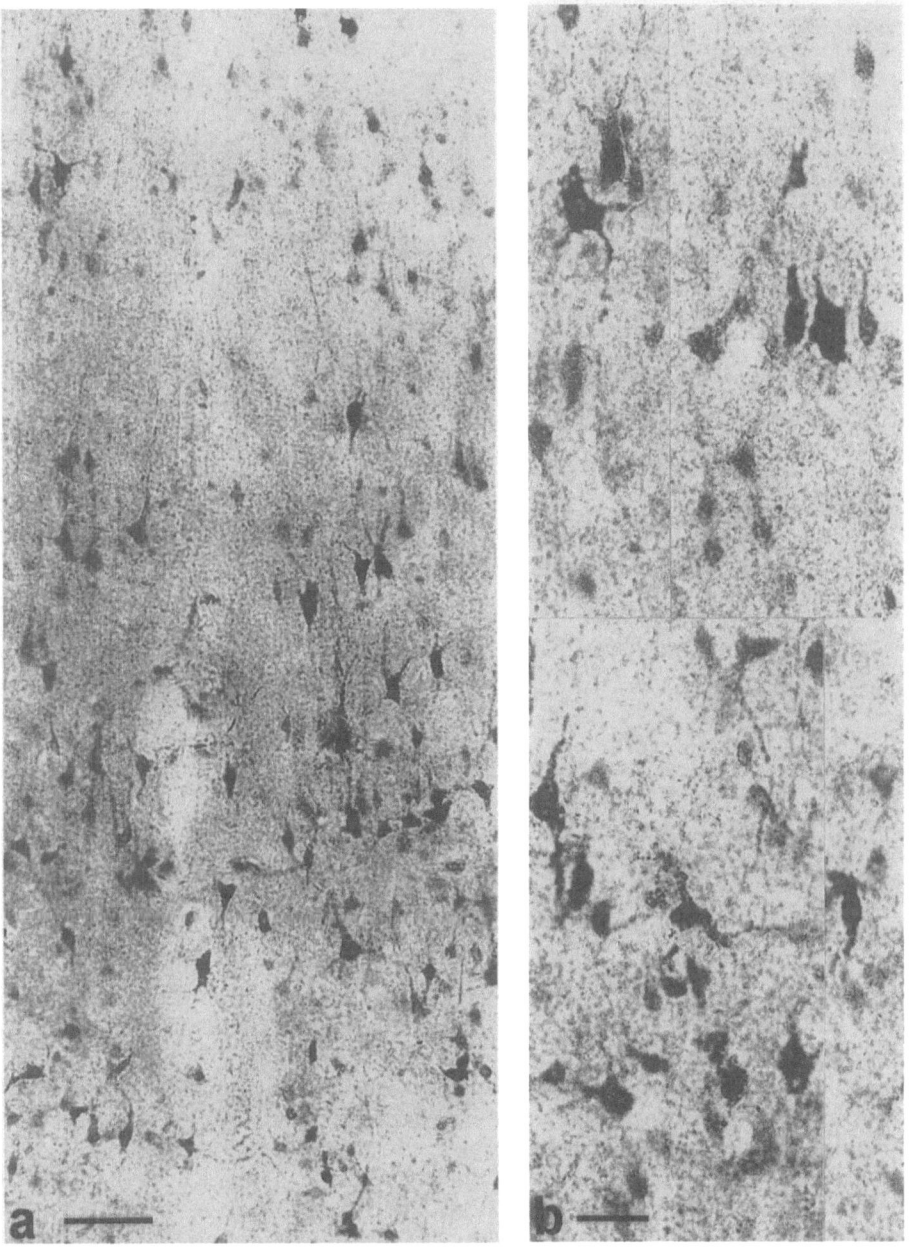

Fig. 6 a, b. CB-ir neurons of the cortical plate (*lower half*) at a survey magnification and **b** high magnification, 28 weeks of gestation. Scale bars: **a**, 50 μm; **b**, 20 μm

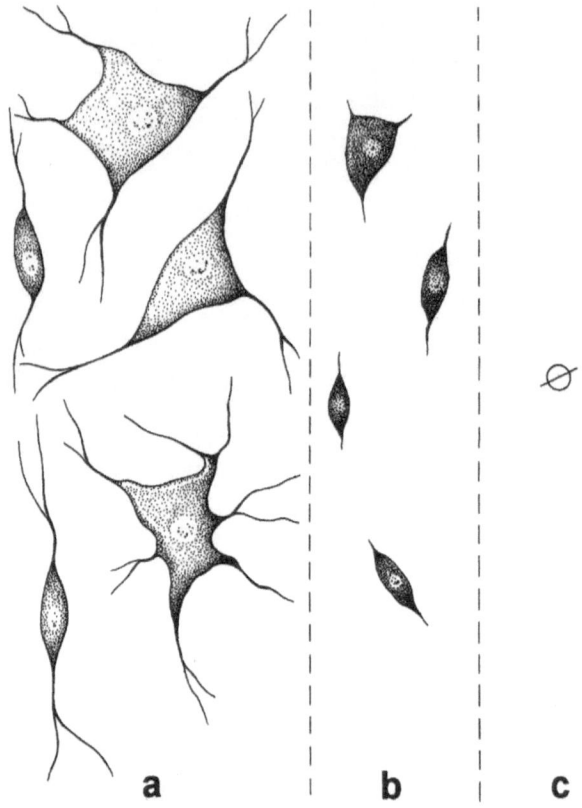

Fig. 7a–c. Alterations in hydrocephalic brains in comparison to controls as seen in PV and CB immunopreparations.
a Control, b severely altered tissue: neurons are distinctly smaller and more intensely immunostained, their number is reduced; only the stems of processed are immunolabeled.
c Extremely altered tissue: no ir structures are observed

a b c

tive neurons that appear shrunken and darkly stained have been described (Wright et al. 1991). Del Bigio (1993) suggested that these reactive neurons correspond to dark neurons. The latter are characterized by a marked condensation of the cytoplasm, which is interpreted as an early reversible alteration in perturbed neurons (Auer and Benveniste 1997). As described in the preceding paragraphs, PV and CB immunopreparations of fetal hydrocephalic brains reveal distinctly smaller and more intensely immunolabeled neurons in the cortical plate than those of control brains. Thus, these neurons may represent dark neurons or a preliminary stage.

In experimental hydrocephalus, a gradual decrease in CB and PV expression with progressive ventricular dilatation has been described (Tashiro et al. 1997). At low degrees of experimental dilatation, a loss of CB and PV immunoreactivity in processes is observed. Moderate degrees cause a reduction in the number of ir nerve cells. In the severest cases of hydrocephalus, no ir nerve cells are found. This progressive loss of CB and PV expression in experimental hydrocephalus corresponds to the aforementioned findings in human fetal hydrocephalus.

With increasing magnitude of ventricular dilatation, a progressive functional impairment of CB and PV synthesis occurs, which finally leads to an exhaustion of these proteins (Tashiro et al. 1997). In both experimental and human fetal hydrocephalus, loss of immunostaining of processes is observed in low degrees of dilatation. Thus, this alteration may be regarded as an initial sign of neuronal damage.

A pronounced difference becomes obvious when comparing the minimal changes that can be observed in Nissl preparations and the severe alterations seen in CaBP immunopreparations. Thus, further investigations on human fetal hydrocephalus using specific histochemical markers are necessary to detect subtle functional changes.

In experimentally induced hydrocephalus in newborn rats, subtle changes of pyramidal neurons have been described: a decrease in the number and length of dendritic branches, an extreme variability in the density of dendritic spines, and frequent appearance of dendritic varicosities (McAllister et al. 1985). The alterations in the pyramidal dendritic trees could result from changes in extrinsic cortical afferents. They may, however, also be ascribed to reduced afferentiation from injured interneurons. As demonstrated in "CaBPs in the Cortical Plate" (Sect. 3.2), cortical interneurons are most likely to be functionally impaired. Pyramidal neurons may, therefore, be damaged through an overexcitation resulting from decreased inhibitory input from the damaged interneurons.

The CaBPs CR, CB, and PV are expressed in largely nonoverlapping subsets of nonpyramidal nerve cells of the cerebral cortex (Baimbridge et al. 1992). These different types of interneurons are differentially susceptible to the effects of fetal hydrocephalus. Even in extremely altered tissue, CR-ir neurons are present, whereas CB- and PV-ir nerve cells are not detectable.

The majority of studies on the effect of hydrocephalus have been performed on laboratory animals. Extrapolation from animals to humans should, however, be made with great caution. In particular, time frames of certain developmental processes cannot be correlated between different species with certainty.

Therefore, investigations on the effects of human fetal hydrocephalus appear necessary. Using antibodies directed against CaBPs, subtle but significant alterations in fetal hydrocephalus could be demonstrated (see also Sects. 2.5 and 8.3). Modifications of sequential developmental events may lead to inappropriate connections, which may account for residual deficits observed after shunting. The evaluation of subtle alterations may have implications for therapeutic decisions, i.e., the optimum timing for surgical intervention.

4 Cerebral Cortex: Molecular Layer (Layer I)

4.1
Description

The molecular layer can be subdivided into two portions: the upper portion (layer Ia) contains large neurons called Cajal-Retzius cells; in the lower portion (layer Ib) horizontally oriented fibers are present. The majority of these fibers represent axons of Cajal-Retzius cells. The packing density of Cajal-Retzius cells is high in the developing brain, whereas in the mature brain these cells are only rarely seen (Retzius 1893; Ramon y Cajal 1911; Marin-Padilla 1998a, 1998b). The function and fate of Cajal-Retzius cells is not fully understood. It has been shown that they synthesize and secrete the glycoprotein reelin. The latter is most likely to act as a stop signal for migrating neurons. Thus, it is involved in the formation of nerve cell layers (D'Arcangelo et al. 1995).

4.2
CaBPs in the Molecular Layer

In CR immunopreparations from brains ranging in age between 21 and 25 weeks of gestation, ir Cajal-Retzius cells in high packing density are seen in layer I (Fig. 8). The perikarya of the ir cells are mainly located in the upper third of layer I, thus forming layer Ia. Horizontal, vertical, and polymorphic ir Cajal-Retzius cells can be distinguished (Fig. 9).

Cajal-Retzius cells also express CB and PV for a short period in the developing monkey cortex (Huntley and Jones 1990).

4.3
Alterations in the Organization of Layer I in Trisomy 22

Although layer I represents a prominent structure in the fetal brain and fulfils important developmental roles, little is known about alterations in this layer in developmental neuropathology. To detect pathological alterations, specific markers to demonstrate the structural elements were used. Cajal-Retzius cells in layer Ia could be visualized with the aid of anti-CR. The axonal plexus in layer Ib could be demonstrated using the antibody SMI 35, which is directed against neurofilament epitopes preferentially found in axons. To demonstrate the formation of synaptic contacts,

Fig. 8. Camera lucida drawings of CR-ir Cajal-Retzius cells of layer Ia, 6th gestational month. Scale bar: 50 μm

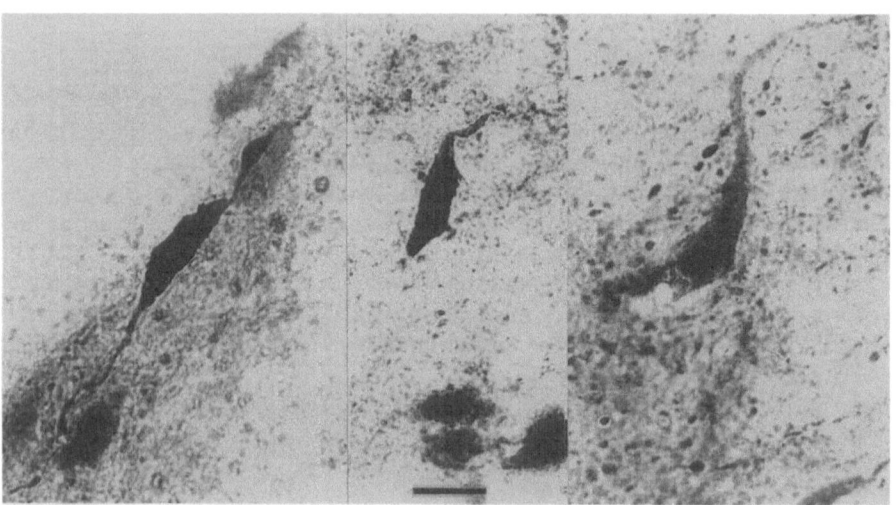

Fig. 9. Cajal-Retzius cells at high magnification, CR immunolabeled, 7th gestational month. Scale bar: 20 μm

anti-GAP (growth-associated protein) 43 and anti-SNAP (synaptosomal-associated protein) 25 were used. Different cases of chromosomal abnormalities (trisomy 18, 21, 22) were investigated in comparison to age-matched controls (Ulfig et al. 1999).

No alterations were observed in layer I in the immunopreparations of trisomy 18 and trisomy 21 in comparison to normal cases (21–25 weeks of gestation): somata of Cajal-Retzius were localized in layer Ia. Layer Ib contained an axonal plexus which was selectively immunostained by SMI 35. Moreover, it harbored GAP 43- and SNAP 25-ir puncta, which were suggestive of synaptogenesis.

In trisomy 22, no alterations were seen in CR immunopreparations. Severe alterations, however, were detected in SMI 35-, GAP 43, and SNAP 25 immunosections. Densely packed SMI 35 ir fibers were observed in layer Ib, arranged in a band-like form. Similarly, GAP 43 and SNAP 25 ir puncta in high packing density were found in layer Ia. Layer Ib was nearly devoid of SMI 35, GAP 43, and SNAP 25 ir structures. Thus, in trisomy 22, layer I displayed malpositioned axonal plexus and a corresponding displacement of synapse-related proteins.

During early cortical development the axons of Cajal-Retzius cells are normally transferred from the depths of the cortical anlage to layer Ib. Obviously, this process is disturbed in trisomy 22.

As indicated by the displacement of synapse-related proteins, the afferents of layer I apparently find their malpositioned targets in layer Ia where the axonal plexus is displaced. These afferents are likely to represent monoaminergic fibers (Verney 1999).

The axonal plexus of Cajal-Retzius cells normally located in layer Ib has been suggested as playing an essential role for the differentiation of pyramidal cells in the cortex (Marin-Padilla 1998b). An interaction taking place between the axonal plexus (of layer Ib) and the apical dendrites of pyramidal cells is essential for the subsequent growth and differentiation of pyramidal cells. This interaction appears to be necessary to establish the polarity of pyramidal cells (Pinto Lord and Caviness 1979; Bayer and Altman 1991; Marin-Padilla 1992). The brains investigated here are derived from fetuses of 23 weeks of gestation; therefore pyramidal cells are not yet differentiated and possible changes in shape or orientation of these cells cannot be examined.

On the whole, it is apparent that the basic functional organization of the molecular layer, i.e., Cajal-Retzius cells, axonal plexus, and specific afferents can be visualized using various markers (anti-CR, SMI 35, anti-GAP 43, and anti-SNAP 25). The latter can also be applied to detect pathological alterations that are not visible in routine preparations, such as Nissl section (Ulfig 1999).

5 Ganglionic Eminence

5.1
Description

Within the neuroepithelium lining, the ventricle proliferation of neuronal and glial precursor cells takes place. In particular during the early and middle period of fetal development, a circumscribed enlargement of the telencephalic proliferative zone is visible. This prominent bulb-like elevation referred to as the ganglionic eminence protrudes into the ventricular cavity (Ulfig et al. 2000c).

5.2
Developmental History

At the beginning of the second gestational month, a thickening of the proliferative zone of the cerebral hemisphere next to the interventricular foramen is seen (Kahle 1969). Shortly after a second protrusion becomes visible. The two elevations, also referred to as the medial and lateral ridge, are separated from each other by a shallow sulcus. The two ridges subsequently gain in size; thus, protrusion into the ventricular cavity becomes more pronounced and growth of the medial ridge causes a narrowing of the interventricular foramen. Thereafter, unification of the two ridges is observed, first in posterior portions of the ganglionic eminence and then in anterior portions (Lammers et al. 1980). With proceeding development, the cerebral hemispheres gain in size. Thus, the ganglionic eminence becomes elongated. As a result of the expansion of the cerebral hemisphere in a curved direction, the ganglionic eminence develops in a C-shaped manner. Consequently, sections of the middle third of the hemisphere pass through the ganglionic eminence, which consists of a superior and inferior part (Ulfig et al. 2000c). The superior part, located laterally in the floor of the central part of the lateral ventricle, borders upon the caudate nucleus. The inferior part, which is found in the roof of the inferior horn of the lateral ventricle, borders upon the amygdala. According to the C-shaped form of the ganglionic eminence, the two parts merge in sections of the occipital lobe.

Only at the end of pregnancy does the ganglionic eminence appear significantly reduced in size, and only its remnants, i.e., thin band-like structures (beneath the amygdala and adjacent to the caudate nucleus), are visible.

5.3
Neuronal Populations Originating
from the Ganglionic Eminence

The ganglionic eminence persists distinctly longer than other proliferative areas and it is known to be the primordium of the striatum (putamen and caudate nucleus), the nucleus accumbens, the amygdala, and the basal nucleus of Meynert (Ulfig et al. 2000c). More recently, substantial evidence has been provided, demonstrating that a significant number of nerve cells in the cerebral cortex are generated in the ganglionic eminence (Parnavelas 2000).

Two principal neuronal types are found in the cerebral cortex, i.e., pyramidal projection neurons and nonpyramidal interneurons. Precursors of pyramidal neurons produced in the telencephalic proliferative zone migrate radially towards the anlage of the cerebral cortex. This migration from the proliferative zone to the target region (i.e., the cortical anlage) is guided by long glial fibers, which are radially oriented. The majority of cortical interneurons are generated in the ganglionic eminence; thus the distance between the place of origin and the final destination is distinctly longer for interneurons than for pyramidal cells. The latter only migrate laterally, i.e., radially. Precursors of interneurons have to leave the ganglionic eminence, migrate tangentially in the intermediate zone (future white matter), and finally migrate radially to reach the cerebral cortex (Tamamaki et al. 1997).

The interneurons of the cerebral cortex represent a heterogeneous neuronal class. Two subpopulations of this class can be defined on account of their expression of CB and CR, respectively.

5.4
Calretinin and Calbindin Immunoreactive Cells
in the Ganglionic Eminence and in the Intermediate Zone

Between 16 and 20 weeks of gestation, a moderate number of CR-ir cells are seen in the center of the ganglionic eminence (Ulfig 2001). In the mantle region, i.e., the periphery, of the ganglionic eminence a large number of CB-ir cells are present. The ir cells of the mantle region mostly display processes that can be traced up to 50 μm. In the intermediate zone, bipolar neuroblasts are found. The appearance of these bipolar cells is typical of migratory neuroblasts with thicker leading and thinner trailing processes. Moreover, the intermediate zone contains tangentially oriented CR-ir fibers (Fig. 10).

From 24 weeks of gestation onwards, the number of CR-ir cells in the ganglionic eminence as well as in the intermediate zone gradually decreases.

Between 16 and 20 weeks of gestation, CB-ir cells are observed in moderate packing density within the mantle region of the ganglionic eminence. From 21 weeks of gestation onwards, a slight increase in number is seen. The CB-ir nerve cells of the mantle region appear larger than the CR-ir cells in the same location (Fig. 11).

CR-ir cells are found in low numbers in the central portion and in high numbers within the mantle zone of the ganglionic eminence. Thus, cells of the ganglionic eminence might be attracted towards the mantle region by a chemoattractant or they

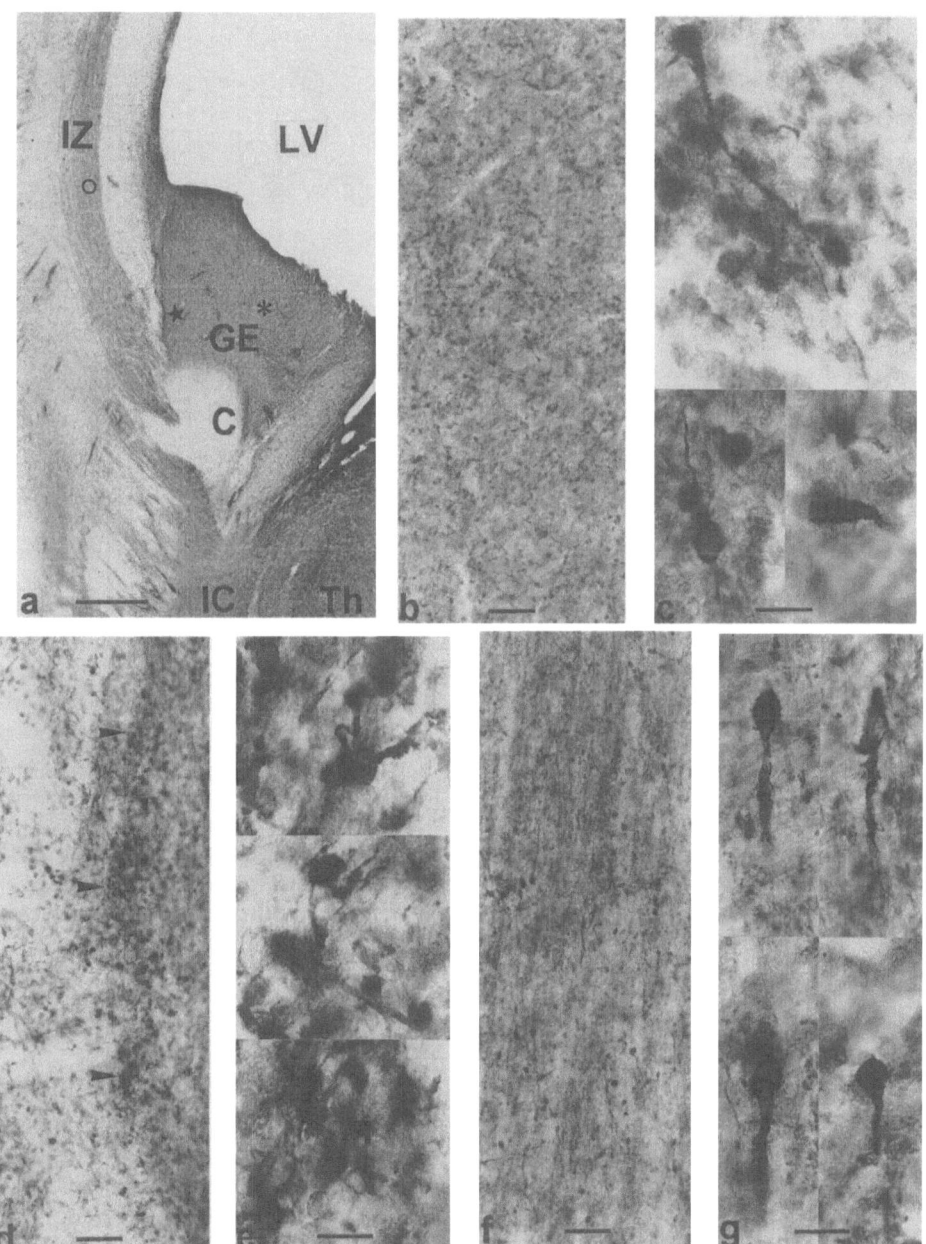

Fig. 10 a–g. a CR-ir structures in the ganglionic eminence (*GE*) and adjacent regions, 16 weeks of gestation. *C*, caudate nucleus; *IC*, internal capsule; *IZ*, intermediate zone; *LV*, lateral ventricle; *Th*, dorsal thalamus. **b** Enlargement taken from the center of the ganglionic eminence, marked by an *asterisk* in **a. c** CR-ir nerve cells in the central portion of the ganglionic eminence. **d** Enlargement taken from the mantle zone of the ganglionic eminence, marked by a *star* in **a. e** CR-ir neurons as seen in the mantle zone of the ganglionic eminence. **f** Enlargement taken from the intermediate zone, marked by an *open circle* in **a. g** CR-ir nerve cells found in the intermediate zone. Scale bars: **a** 1 mm; **b, d, f**, 50 μm; **c, e, g**, 10 μm. (From Ulfig 2001, with permission)

31

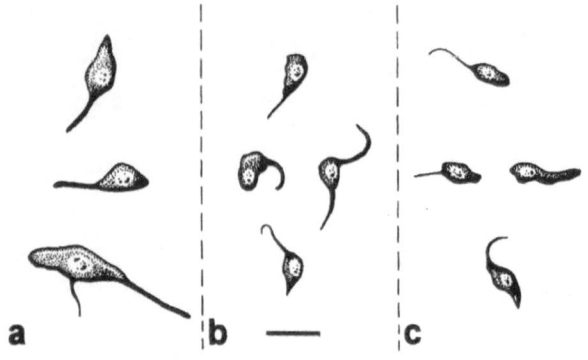

Fig. 11 a–c. Camera lucida drawings of a CB-ir neurons in the ganglionic eminence, b CR-ir neurons in the ganglionic eminence, c CR-ir neurons in the intermediate zone. 24 weeks of gestation. Scale bar: 10 μm. (After Ulfig 2001 with permission)

might migrate in random directions and reach the mantle region by coincidence (Ulfig 2000a).

These observations indicate that CR-ir neuroblasts leave the ganglionic eminence, then cross the corticostrial boundary and follow tangential migratory routes on their way to the cerebral cortex, where they contribute to the class of inhibitory interneurons. It has been postulated that axons provide a substratum for this tangential, i.e., nonradial, neuronal migration (Tamamaki et al. 1997). In line with this notion, an association of tangentially oriented CR-ir cells and similarly running CR-ir fiber has been observed (see Chap. 4, "Neuronal Populations Originating from the Ganglionic Eminence").

Another finding demonstrating tangential migration from the ganglionic eminence is worth mentioning. Precursor neurons of the ganglionic eminence have been shown to migrate across the telencephalodiencephalic boundary and finally settle within dorsal thalamic nuclei. On their way to the dorsal thalamus, these neurons migrate through a distinct transient structure, referred to as the gangliothalamic body (Rakic and Sidman 1969; Letinic and Kostovic 1997; Ulfig et al. 2000b).

5.5
The Ganglionic Eminence as an Intermediate Target

In SNAP-25 immunopreparations from brains between 19 and 24 weeks of gestation, ir fibers are observed in the intermediate zone and in the internal capsule. The ir fibers run in parallel to each other. At the ventricular edge, they reveal curvature and many of the fibers appear to be directed towards the lateral margin of the ganglionic eminence. Within this lateral margin, outstandingly intense SNAP-25 immunolabeling is observed. The remaining portion of the ganglionic eminence and the subjacent caudate nucleus are devoid of SNAP-25 ir structures. At high magnification, a pericellular immunostaining pattern is seen in the mantle region.

In preparations double-immunostained with anti-SNAP-25 and anti-CR, single-labeled CR-ir nerve cells appear brown, single-labeled SNAP-25-ir puncta or fibers appear blue, and double-labeled nerve cells (CR-ir with pericellular SNAP-ir puncta) appear black. In the mantle region of the ganglionic eminence, a high number of double-labeled nerve cells are visible; they are embedded in SNAP-25-ir puncta. Cells of the central part of the ganglionic eminence only reveal CR immunostaining.

The punctate (pericellular) SNAP-25 immunostaining pattern is known to reflect the localization of SNAP-25 in presynaptic terminals. Fibrous SNAP-25 immunolabeling indicates an accumulation of SNAP-25 within axons.

The CR- and CB-ir cells in the mantle region of the ganglionic eminence are likely to fulfill an important role in axonal pathfinding. In coculture experiments, an oriented growth of cortical axons towards the ganglionic eminence has been observed (Metin and Godement 1996). Thus, the cells of the mantle region are likely to serve as an intermediate target for outgrowing axons.

Outgrowing axons may make characteristic changes in direction as they reach a specific site. Cells of such a specific area may represent an intermediate target and act as guidepost cells for a developing projection. On their way to their target, thalamo-cortical fibers seem to pause within the mantle region of the ganglionic eminence. This part of the ganglionic eminence is known to contain postmitotic neurons, which probably correspond to the CR- or CB-ir cells described in Sect. 5.4. As described in this section, antibodies directed against SNAP-25 have been used to demonstrate corticofugal axons reaching the ganglionic eminence (Ulfig et al. 2000d). The SNAP-25-ir fibers make a turn to reach the lateral mantle region of the ganglionic eminence. The fibers are likely to reside here prior to growing further towards the thalamus.

The cell bodies of the mantle zone of the ganglionic eminence have been shown to receive axonal appositions from growing cortical axons (Metin and Godement 1996). The existence of axocellular appositions in the human fetal brain is shown by the above-mentioned demonstration of CR- and SNAP-25 double-labeled cell bodies in the mantle region.

The growth of fibers towards the ganglionic eminence (as an intermediate target) is most probably mediated by chemoattractants which are released by the cells of the intermediate target. A likely candidate acting as a chemoattractant has been shown to be netrin-1; an increasing gradient of netrin-1 may attract outgrowing corticofugal axons towards the mantle region of the ganglionic eminence (Metin et al. 1997). Another mechanism underlying the orientation of corticofugal fibers towards the ganglionic eminence is conceivable. During early development the ganglionic eminence is likely to provide a projection towards the cerebral cortex. Thus, at least some early corticofugal axons might navigate their way in the opposite direction towards the ganglionic eminence along a pathway formed by pioneering axons of cells of the ganglionic eminence (Metin and Godement 1996; Molnar 1998; Ulfig et al. 2000a).

5.6
Significance of the Ganglionic Eminence in Developmental Neuropathology

Hemorrhage in the ganglionic eminence is a major CNS complication in prematurely born infants. In particular, very young premature infants display a distinctly high incidence of hemorrhage in the ganglionic eminence (Volpe 1995). Hemorrhage is known to cause severe handicap in the infant; motor as well as cognitive developmental outcome may be impaired. So far, the mechanisms underlying cognitive deficits are only poorly understood. An impairment of these functions of the ganglionic eminence, i.e., an intermediate target and major source of cortical interneurons, may at least in part account for intellectual deficits in the infant. As demonstrated above, antisera against CR and CB are appropriate tools to investigate the functional organization of the ganglionic eminence in cases of hemorrhage.

6 Striatum

6.1
Description

The striatum is composed of two distinct compartments: the patch compartment (striosomes) and matrix. Patches have been shown to be neurochemically different from the matrix (Graybiel et al. 1981; Graybiel 1990; Liu and Graybiel 1992b). Moreover, the two compartments show differences in afferent and efferent connections.

The striatal neurons are generated in the ganglionic eminence. A relationship between the generation of a neuron and its final location within one of two striatal compartments has been shown in the rat brain. Neurons destined for the patch compartment become postmitotic distinctly earlier than those that become matrix neurons. Thus, the nerve cells of the two compartments are generated during mainly nonoverlapping developmental periods (van der Kooy 1987; Marchand and Lajoie 1986).

In the adult brain, the patches show very low acetylcholinesterase (AChE) activity whereas the surrounding matrix is characterized by high AChE-activity. During fetal and neonatal development, however, the patches show intense AChE-labeling, whereas the matrix appears distinctly lighter (Graybiel and Ragsdale 1980; Graybiel 1990). The AChE-rich patches in the developing brain have been shown to contain a high number of dopaminergic fiber terminals. These fibers arise from the substantia nigra (Voorn et al. 1988).

During postnatal development, progressive changes take place: an increase in matrix- and a decrease in patch-AChE-staining lead to the completely reversed distribution pattern of AChE activity in the adult striatum.

6.2
Calbindin Expression in the Developing Striatum

Between 16 and 40 weeks of gestation, the patch compartment stands out clearly in CB immunopreparations; it is characterized by intense neuropil labeling (Fig. 12). The surrounding matrix appears distinctly lighter as neuropil is not immunostained in this compartment. Instead, the matrix contains CB-ir cell bodies. Two neuronal types of CB-ir nerve cells have been distinguished (Fig. 13):
1. Small to medium-sized neurons are frequently encountered in the striatal matrix. These weakly immunostained nerve cells are seen in various stages of maturation.

Fig. 12. Distribution of CB-ir structures
in the striatum, 7th gestational month.
Larger dots represent ir neurons in the matrix,
small dots diffuse (neuropil) immunostaining
in the patches

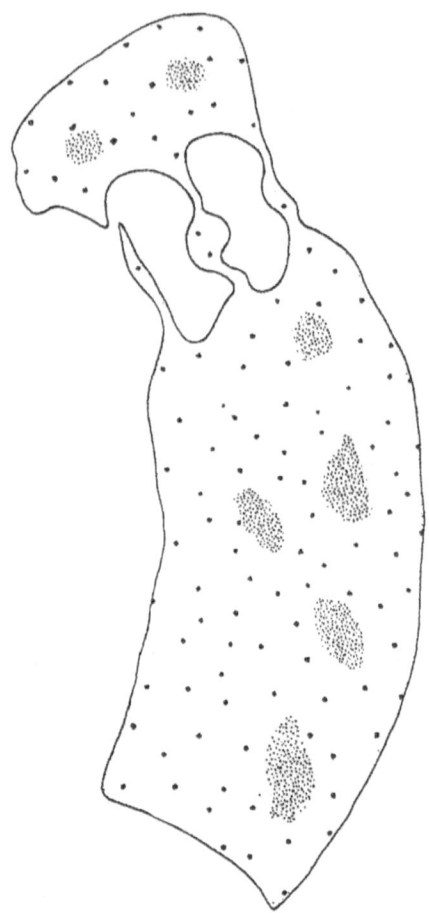

2. Large neurons are more intensely immunolabeled and display long ramified processes. The large neurons are less numerous than the small to medium-sized neurons. In the postnatal brain, this neuronal type is no longer present.

Letinic and Kostovic (1996b) demonstrated that the patches rich in CB-ir neuropil closely correspond to AChE-rich patches in serially adjoining sections of prenatal brains. Thus, anti-CB can be used as a reliable marker of the patch compartment in human fetal brains.

In postnatal brains, a distinct changeover in the distribution pattern of CB-ir neuropil is observed. In the brain of an 18-month-old infant, patches appear lightly immunostained, whereas the surrounding matrix reveals dense immunolabeling. Similarly, the pattern of AChE-reactivity changes. Thus, in serially adjoining sections,

Fig. 13a, b. CB-ir neurons as seen in the striatal matrix, 8th gestational month. **a** Intensely immunostained neurons. **b** Weakly immunostained neurons. Scale bar: 20 µm

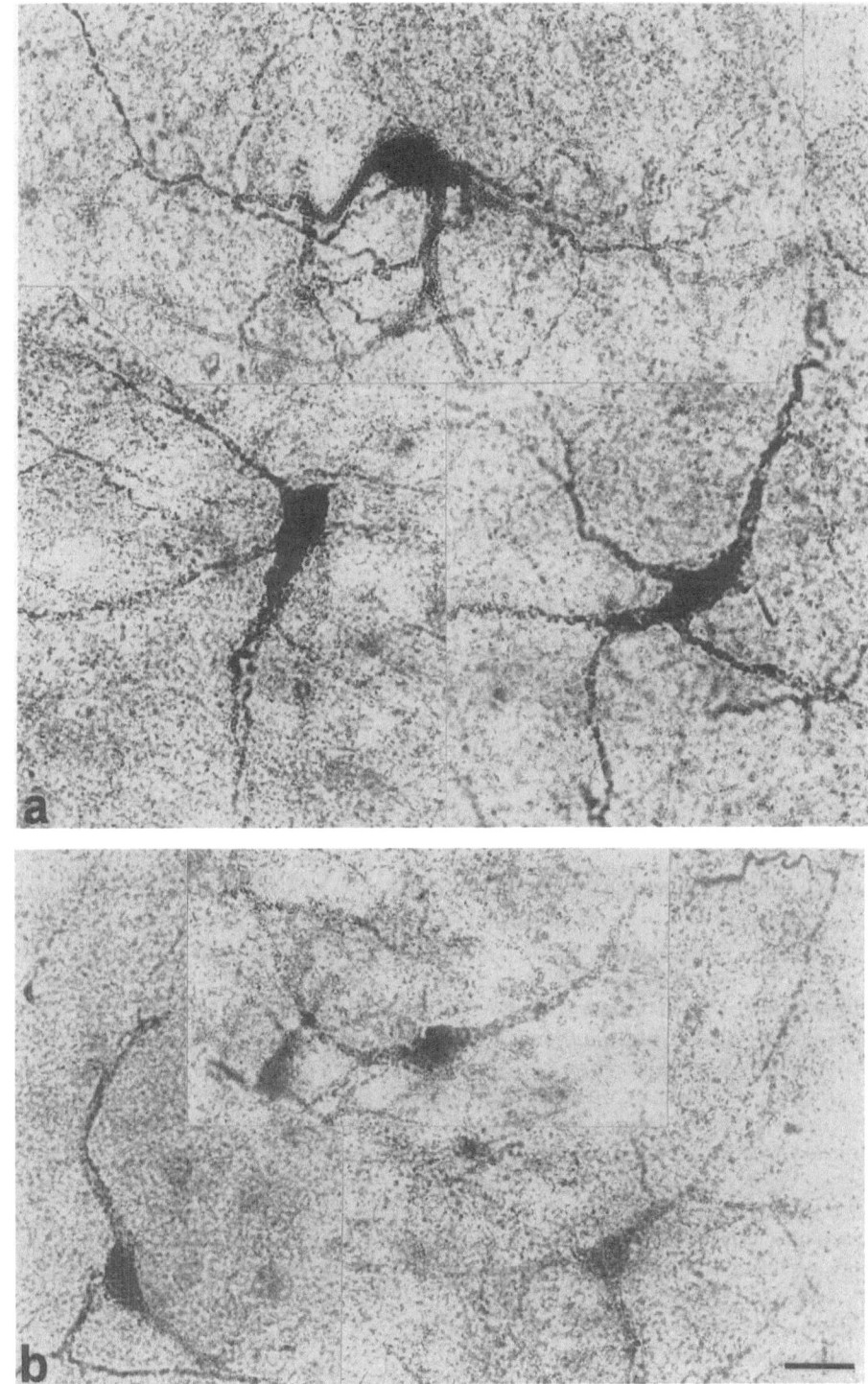

37

patches that are only lightly CB immunolabeled closely match AChE-poor patches. These staining patterns observed in the 18-month-old infant correspond to those found in adult brains.

In the rat brain, neurons of the patch compartment become postmitotic between E12 and E17, and those of the matrix between E17 and E20 (Marchand and Lajoie 1986). The first generated neurons later forming the patch compartment are initially evenly distributed within the striatal anlage. These neurons appear to stay clustered together when matrix neurons migrate into the striatum. Matrix neurons do not tend to be clustered but settle well spaced apart in the matrix between the patches (Song and Harlan 1994).

The earliest fibers that reach the striatum originate in the substantia nigra and are dopaminergic. Until E19, the dopaminergic terminals are evenly distributed within the rat striatum (Voorn et al. 1988). Subsequently, a distinct transition from an even to a patchy distribution takes place: dopaminergic innervation is mainly restricted to the patch compartment. It has been postulated that CB immunoreactivity and AChE reactivity in the patch compartment are contained in nigrostriatal fibers, which form the dopamine islands in the developing striatum (Graybiel 1984; Letinic and Kostovic 1996a).

The mechanisms underlying the processes of nerve cell and fiber coalescing within the patch compartment have so far not been completely understood. It has been suggested that cell surface molecules account for adhesion of nerve cells and their clustering in patches (Krushel et al. 1989). Furthermore, the borders between the two striatal compartments are enriched with certain extracellular matrix molecules such as tenascin (Steindler et al. 1988). The latter is known to restrict the areas where axons can grow (Faissner and Schachner 1995). Thus, constraining molecules may be involved in the constriction of fibers within the patch compartment (Charvet et al. 1998).

6.3
Correlation of Calbindin Immunostaining and Expression of Synapse-Related Proteins in the Striatal Compartment of the Human Developing Brain

The two striatal compartments are known to receive different inputs which are established sequentially during ontogenesis. Therefore, it would be interesting to study synaptogenesis to see the maturation of the patches and the matrix (Ulfig et al. 2000e, 2001a). Antibodies directed against synaptophysin have been shown to be useful tools for studying synaptogenesis (Masliah et al. 1990; Ulfig and Chan, 2002b). Synaptophysin is a 38-kDa synaptic-vesical-associated protein, which is widely distributed within the CNS (Jahn et al. 1985). It is involved in synaptic vesicle fusion with terminal membrane of the axon and subsequent release of the neurotransmitter.

In serially adjacent immunopreparations from brains ranging in age from 21 to 25 weeks of gestation, the distribution of CB- and synaptophysin-ir structures has been compared (Fig. 14). In CB immunosections, the patches stand out clearly due to their intense diffuse (neuropil) immunolabeling; the matrix only contains ir perikarya and their dendritic trees. In serially adjoining synaptophysin immunosections, the patches exhibit prominent diffuse labeling. In the matrix compartment, only very few

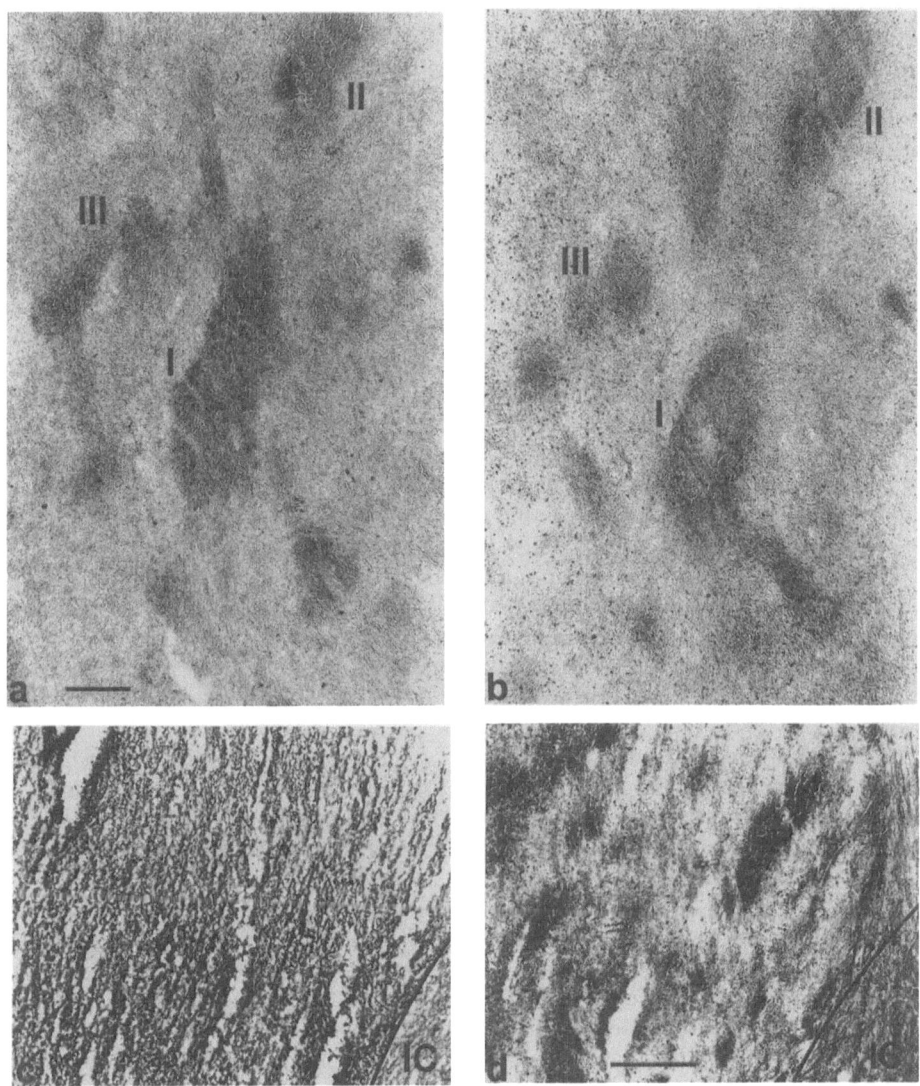

Fig. 14. a Patchy distribution of synaptophysin-ir structures in the putamen, 25 weeks of gestation. b Serially adjoining section immunostained with anti-CB. Note the similar arrangement and configuration of diffusely immunostained patches (*I–III*). c Homogeneous distribution of synaptophysin-ir structures in the putamen, 31 weeks of gestation. d Serially adjoining CB immunosection displaying a patchy distribution of CB-ir puncta. *IC*, internal capsule. Scale bars: a and b, 100 μm; c and d, 200 μm. (From Ulfig et al. 2000e, with permission)

and weak ir puncta are present. The configuration and arrangement of synaptophysin-ir patches closely match those of CB-ir patches.

Between 28 and 32 weeks of gestation, the striatum exhibits a very similar distribution of CB-ir structures, as described for the age group 21–27 weeks of gestation. In contrast, serially adjoining synaptophysin immunosections (between 28 and 32 weeks of gestation) reveal a homogeneous immunolabeling within the striatum; no differences between the patches and the matrix are seen.

The A Kinase Anchoring Protein (AKAP) 79 is enriched in postsynaptic densities, which represent a network of proteins found on the internal surface of excitatory synapses (Klauck and Scott 1995). Regulatory proteins may be tethered near or within postsynaptic densities by AKAP 79 (Klauck et al. 1996). The level of rat AKAP 150, which is related to the human AKAP 79, varies considerably among brain regions and nerve cell types (Rubin 1994; Coghlan et al. 1995). It has been assumed that AKAP 150 may be attributed to pathways containing a definite neurotransmitter such as dopamine (Glantz et al. 1992). Therefore, the correlation between AKAP 79 and CB expression has been investigated in the human fetal striatum. In both CB as well as AKAP 79 immunopreparations, the patches stand out conspicuously due to their intense diffuse labeling at 27 weeks of gestation. The neuropil of the surrounding matrix displays only weak immunolabeling (Fig. 15). The arrangement and configuration of AKAP 79-ir patches closely correspond to those of CB-ir patches in serially adjoining sections. This distribution pattern of AKAP 79-ir structures within the striatum does not change during proceeding development.

Using CB immunohistochemistry, a transition from a patchy to matrix neuropil labeling during postnatal development is observed. This reorganization can most likely be correlated with the sequential arrival of different afferent fibers. The patchy neuropil labeling observed in the fetal and newborn brains may reflect the early dopaminergic nigrostriatal projection to the patch compartment. Later these dopaminergic fibers project to the matrix and the patches.

In the postnatal and adult striatum, however, CB ir neuropil is only observed in the matrix. This pattern can be explained when assuming that CB is restricted to nigrostriatal fibers projecting only to the matrix. In line with this notion, CB has been shown to be selectively expressed in nigral projections to the striatal matrix (Gerfen et al. 1985).

When comparing CB and synaptophysin-immunostaining patterns in the striatum from the 8th gestational month onwards, two differences are observed. One difference concerns the period of time in which the transformation of staining patterns takes place. An even distribution of synaptophysin-ir puncta within the striatum is detected at the beginning of the 8th month of gestation, whereas the patchy CB labeling changes only postnatally. Mainly two reasons may account for this difference. First, the synaptic vesicle protein may be present before synaptogenesis (Leclerc et al. 1989; Sarthy and Bacon 1985; Bergmann et al. 1991; Reisert and Gratzl 1993). Second, synaptophysin is the predominant synaptic vesicle protein in other afferent fiber systems, i.e., the thalamostriatal and the corticostriatal projections. The arrival of these fibers is most likely to contribute to the transformation of the patchy immunolabeling (Ovtscharoff et al. 1990).

The second difference between CB- and synaptophysin immunopreparations concerns the distribution patterns after the transformation of the patchy immunolabeling. CB-positive fibers are only found in the matrix, whereas synaptophysin-ir puncta are seen homogeneously distributed within the striatal compartments. This difference may readily be explained by the fact that synaptophysin is found in various striatal afferents (see this section, above), whereas CB is likely to be restricted to nigrostriatal fibers.

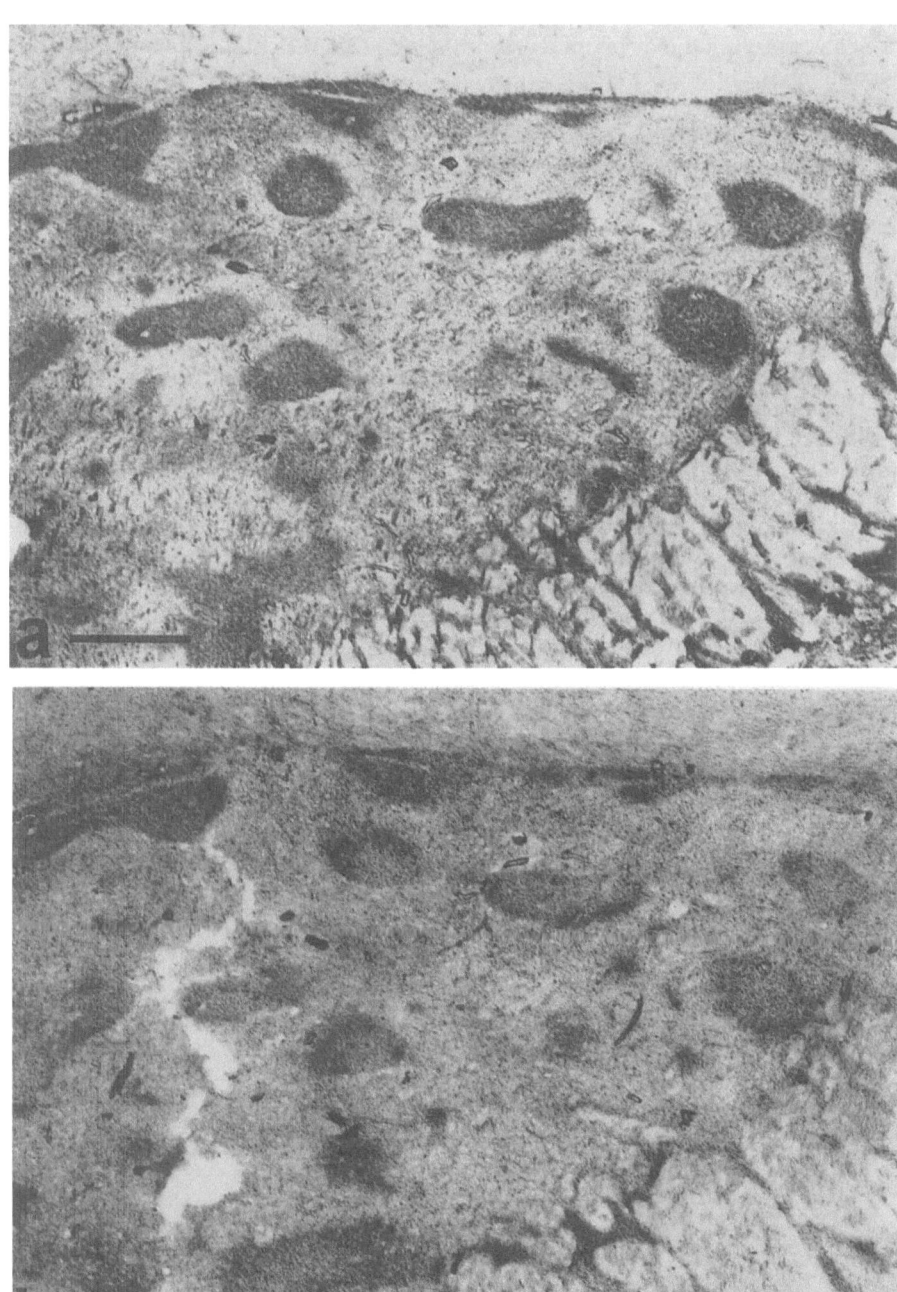

Fig. 15. a Distribution of AKAP-79-ir structures in the putamen. **b** Serially adjoining section immunostained with anti-CB. Note the similar arrangement of diffuse immunolabeling in the patches in **a** and **b**. 27 weeks of gestation. Scale bar: 1 cm. (From Ulfig et al. 2001a, Karger Basel, with permission)

When comparing the developmental profiles of AKAP 79 and CB expression, it is obvious that the mature pattern of AKAP 79 expression is detectable already at 27 weeks of gestation. Thus the changeover as observed in CB distribution in the postnatal brain is not reflected by the expression pattern of AKAP 79. This discrepancy has not yet been explained sufficiently. It cannot, however, be ruled out that AKAP 79 is expressed at mature levels before the mature pattern of striatal input is developed.

6.4
Significance of the Immunohistochemical Findings in the Striatum for Developmental Neuropathology

The different distribution patterns of CB-, synaptophysin-, or AKAP 79-ir structures at various developmental stages provide a basis for further investigations in the field of developmental neuropathology.

Hypoxic-ischemic brain injury is a leading cause of neurological morbidity in premature as well as full-term infants. The striatum is one of the major sites of predilection for neuronal damage in hypoxic-ischemic encephalopathy. The incidence of this neuronal injury is often underestimated. It is characterized by extrapyramidal syndromes such as choreoathetosis or dystonia, which occur 1 year and often much later after lesioning (Volpe 1995). The morphological substrate for these clinical observations has so far not been determined.

7 Amygdala

7.1
Description

The amygdaloid complex is localized in the depth of the temporal lobe. It consists of a series of nuclei, each of which displays distinct cyto- and chemoarchitectural and connectional characteristics (Amaral et al. 1992; Fallon and Ciofi 1992). The amygdaloid body can be divided into basolateral and corticomedial nuclei. The basolateral part corresponds to the deep nuclei of Amaral et al. (1992); it encompasses the lateral, basal, accessory basal and paralaminar nuclei. The corticomedial part, corresponding to the superficial nuclei, consists of the anterior cortical, posterior cortical, medial and central nuclei, and the periamygdaloid cortex. The remaining nuclei and areas which cannot be assigned to one of the aforementioned parts are the anterior amygdaloid area, the amygdalohippocampal area, and intercalated nuclei.

The corticomedial nuclei are mainly involved in autonomic and endocrine functions via connections with the hypothalamus and the brain stem. The basolateral nuclei project to many iso- and allocortical areas and receive afferents from association cortices. Intrinsic connections provide a flow of information from the basolateral to the corticomedial nuclei. On the whole, the basolateral complex represents an important link between cortical input and amygdaloid output to cortical and – via the corticomedial nuclei – subcortical areas. On account of these connections, the basolateral nuclei participate in complex behavioral patterns such as defense or aggression, learning, and memory (Aggleton 1993; Swanson and Petrovich 1998).

Cytoarchitectonic studies on the human embryonic brain have shown that amygdaloid nuclei first become visible in the 8th week of gestation and that the corticomedial nuclei occur earlier than the basolateral (Macchi 1951; Humphrey 1968).

7.2
Transient Architectonic Features of the Amygdala

During fetal development, the basolateral amygdaloid nuclei directly border the inferior part of the ganglionic eminence. In this border area, distinct architectonic features are visible, mainly in the 5th and 6th gestational month (Ulfig et al. 1998c).

In Nissl-stained preparations, the lateral nucleus displays elongated column-like cell aggregations in its inferior and lateral portions. These broad cell-dense columns are separated from each other by narrow cell-sparse zones. The latter extend into the ganglionic eminence; thus the cell columns continuously extend into the ganglionic eminence.

Columnar cell clusters are only slightly developed in the basal nucleus, whereas the accessory basal nucleus reveals thin columnar cell clusters which continuously extend into the ganglionic eminence.

These cell-dense columns most likely reflect clustered migrational routes from the ganglionic eminence (proliferative zone) to the basolateral amygdaloid nuclei (Nikolic and Kostovic 1986). In line with this notion, the cell-dense columns have been shown to harbor a large number of vimentin-ir radial glial fibers (Ulfig et al. 1998c). The latter are known to play a crucial role in the migration of precursor cells from their proliferative areas to their target positions (Rakic 1990).

In the 7th gestational month, the cell-dense columns of the basolateral nuclei loose their contacts with the ganglionic eminence. Columnar cell arrangements are no longer observed in the ganglionic eminence. In the 8th–9th gestational month, no columnar cell clusters are detectable in the basolateral nuclei. A cell-free capsule surrounds the basolateral nuclei and clearly separates them from the remnants of the ganglionic eminence. In accordance with these findings, vimentin immunopreparations show a remarkable reduction in radial glial fibers.

On the whole, changing characteristics reflecting reorganization of the cytoarchitecture are observed at the interface of the ganglionic eminence and the amygdala. These transient features should be taken into consideration when investigating distribution patterns of neuroactive substances such as CaBPs (see Sect. 7.3.1).

7.3
Nerve Cell Types in the Fetal Amygdala as Seen in Calbindin and Calretinin Immunopreparations

CB and CR show a widespread distribution within the fetal amygdala. These two CaBPs are expressed in distinct subpopulations of neurons, which may, therefore, be distinguished by specific calcium-dependent processes (Setzer and Ulfig 1999).

7.3.1
Neurons in the 5th–6th Gestational Month

In the 5th and 6th gestational months, small to medium-sized neurons can be seen in the corticomedial as well as in the basolateral parts (Figs. 16–18).

The corticomedial nuclei harbor CB- and CR-ir with medium-sized somata from which two to four dendrites emerge. The latter branch is some distance from the soma. In the basolateral nuclei, small ovoid CB- and CR-ir perikarya, which give rise to one or two unbranched processes, are observed (Figs. 19–20). The number of CR-ir nerve cells is distinctly higher than that of CB-ir neurons (Figs. 17, 18).

Moreover, a moderate number of small bipolar neurons oriented perpendicularly to the ganglionic eminence is found. The form, size, and orientation of these neurons suggest that they most probably represent migrating neurons. When analyzing serially adjoining Nissl sections, it is obvious that these presumably migrating neurons are

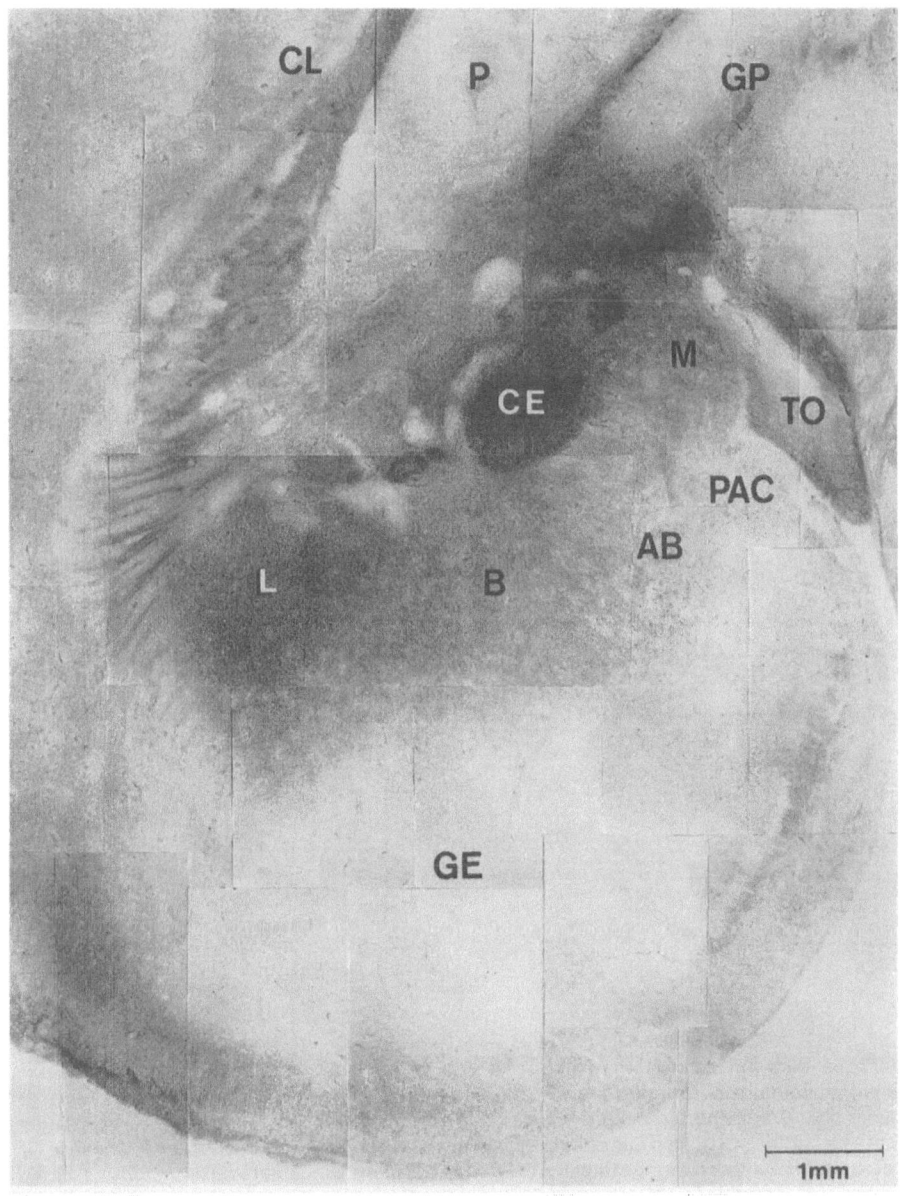

Fig. 16. Distribution of CB-ir structures within the amygdala in the 5th–6th gestational month. For abbreviations, see Fig. 17. Scale bar: 1 mm. (After Setzer and Ulfig 1999, with permission)

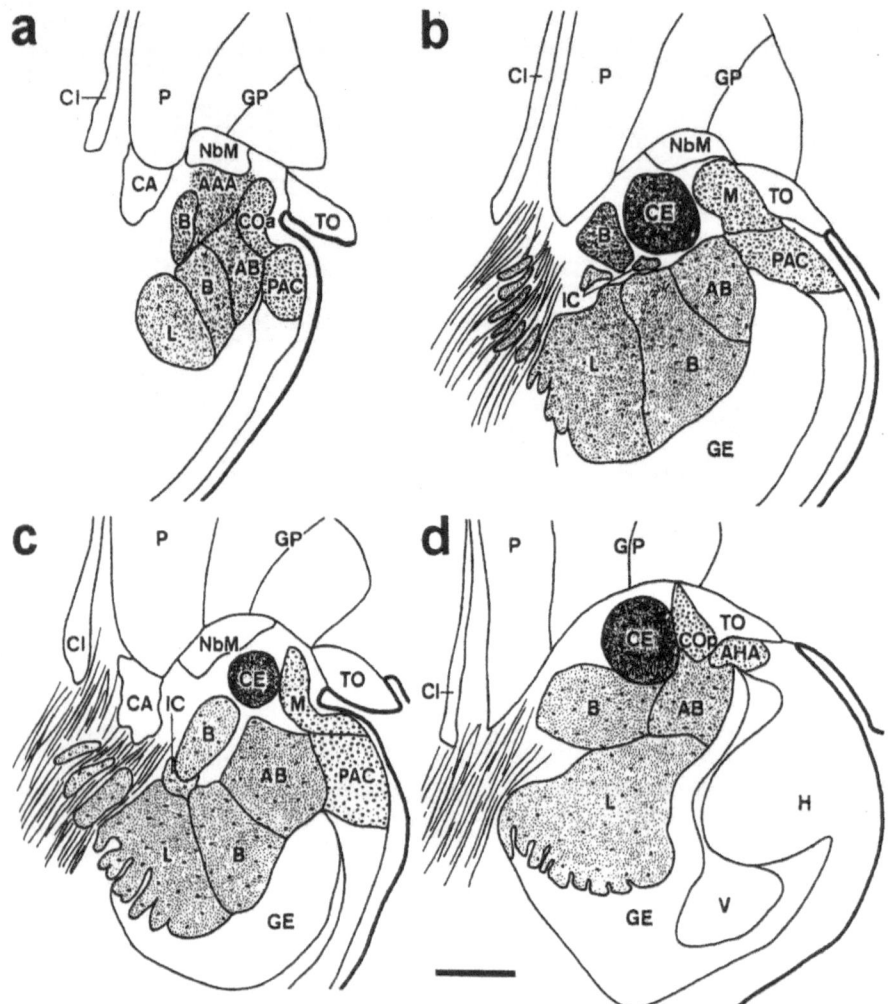

Fig. 17a–d. Distribution of CB-ir structures within the amygdala in the 5th–6th gestational month. The amygdaloid nuclei are schematically drawn in frontal section from **a** anterior to **d** posterior. The number of *large dots* represents the number of ir neurons; the intensity of *small dots* represents the intensity of diffuse (neuropil) immunolabeling. *AAA*, anterior amygdaloid area; *AB*, accessory basal nucleus; *AHA*, amygdalohippocampal area; *B*, basal nucleus; *CA*, anterior commissure; *CE*, central nucleus; *Cl*, claustrum; *Coa*, anterior cortical nucleus; *Cop*, posterior cortical nucleus; *GE*, ganglionic eminence; *GP*, globus pallidus; *H*, hippocampus; *IC*, intercalate nuclei; *L*, lateral nucleus; *M*, medial nucleus; *NbM*, basal nucleus of Meynert; *P*, putamen; *PAC*, periamygdaloid cortex; *PIR*, piriforme cortex; *PL*, paralaminar nucleus; *TO*, optic tract; *V*, ventricle. Scale bar: 1 mm. (After Setzer and Ulfig 1999, with permission)

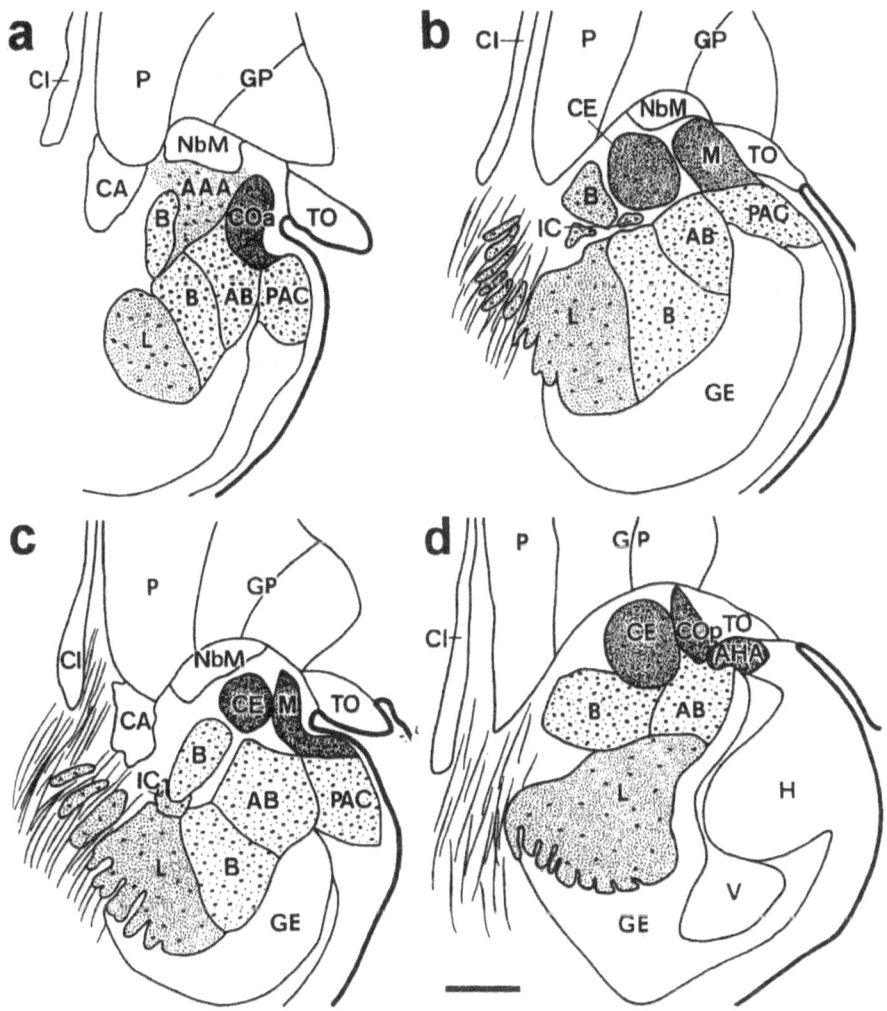

Fig. 18a–d. Distribution of CR-ir structures within the amygdala in the 5th–6th gestational month. For further details, see Fig. 17. Scale bar: 1 mm. (After Setzer and Ulfig 1999, with permission)

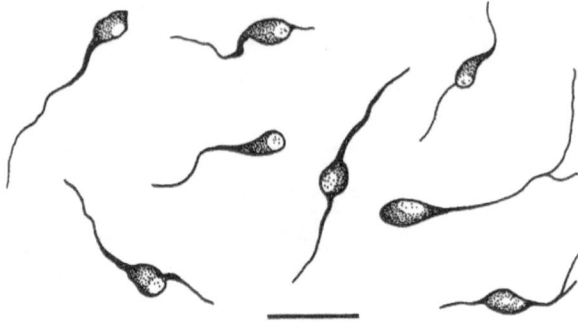

Fig. 19. Camera lucida drawings of CB-ir neurons within the amygdala in the 5th gestational month. Scale bar: 25 μm. (After Setzer and Ulfig 1999, with permission)

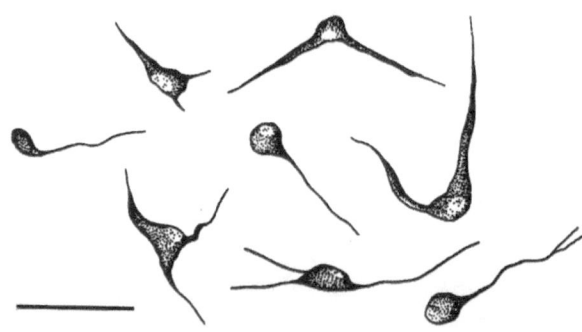

Fig. 20. Camera lucida drawings of CR-ir neurons within the amygdala in the 5th gestational month. Scale bar: 25 μm. (After Setzer and Ulfig 1999, with permission)

located within columnar cell clusters. The latter most probably represent migratory routes (see Sect. 7.2).

Calcium-binding proteins are most likely to be of importance in neuronal migration because calcium ions have been shown to regulate cell movement (Frassoni et al. 1998). In vitro experiments could demonstrate that the amplitude and the frequency of intracellular calcium fluctuations are linked to cell movement (Komuro and Rakic 1996).

Apart from being expressed in migrating neurons, CR and CB are obviously found in immature neurons after they settle into target regions. Evidence has been provided that calcium fluctuations play an important role in the maturation of ion channels and the occurrence of neurotransmitters (Spitzer 1994). Moreover, levels of intracellular calcium are involved in regulating neurite outgrowth, growth cone motility, and expression of neurotransmitter receptors (Gomez et al. 1995; Kater et al. 1988). CaBPs may regulate the levels of intracellular calcium through their ability to buffer calcium.

7.3.2
Neuronal Types Observed in the 8th–9th Gestational Month

From the 8th gestational month onwards different neuronal types can be distinguished in the immunopreparations (Figs. 21, 22).

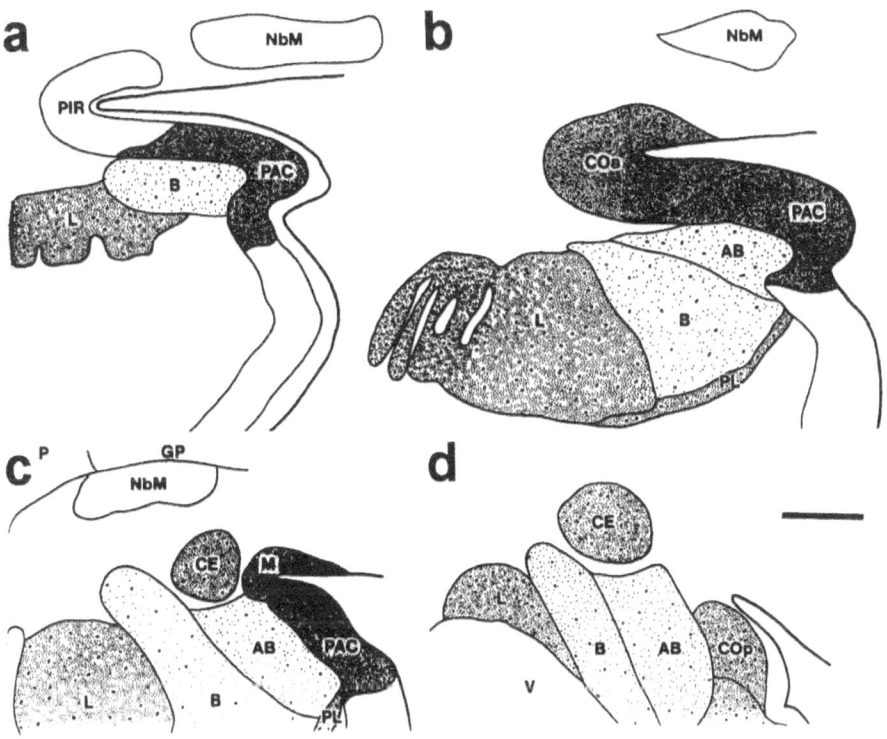

Fig. 21. Distribution of CB-ir structures within the amygdala in the 8th–9th gestational month. For further details, see Fig. 17. Scale bar: 1 mm. (After Setzer and Ulfig 1999, with permission)

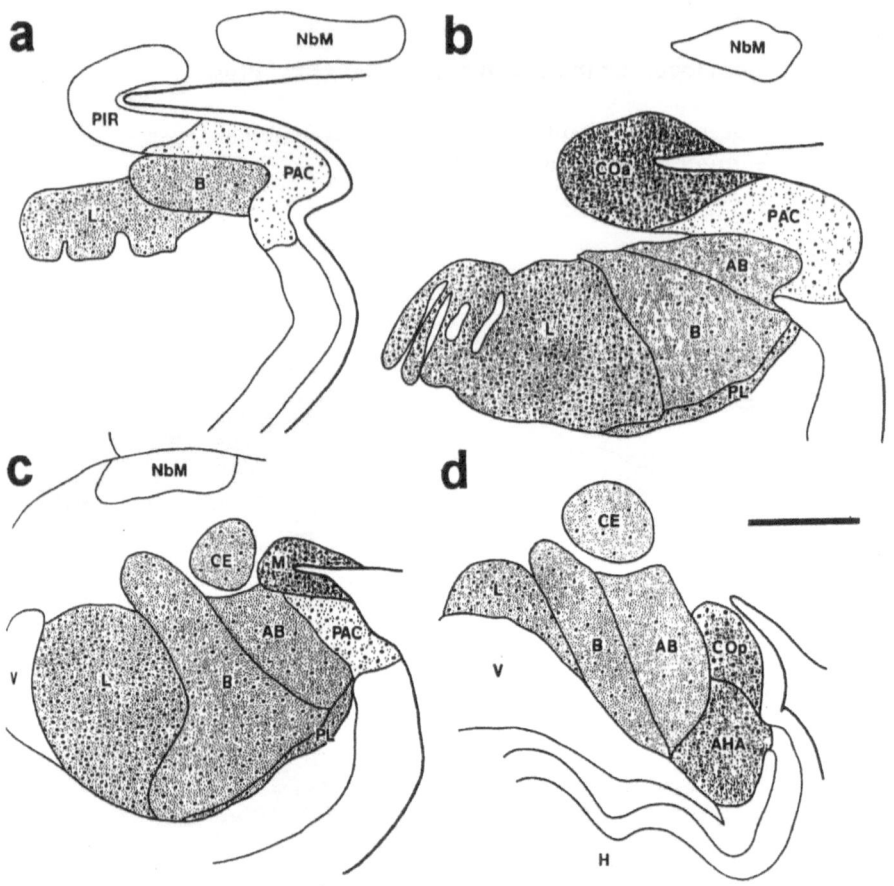

Fig. 22. Distribution of CR-ir structures within the amygdala in the 8th–9th gestational month. For further details, see Fig. 17. Scale bar: 1 mm. (After Setzer and Ulfig 1999, with permission)

7.3.2.1
Presumed Pyramidal Neurons (CB- and CR-ir)

From the base of the pyramid-shaped soma, thin dendrites emerge. At the opposite site, the soma merges into a stout main dendrite, which generates several branches. Occasionally, one of the basal dendrites shows a prominent stoutness and gives off several branches close to the soma. From the base of pyramidal cells, the axon emerges. The latter often bends near the soma. The axon may also emerge from the stem of a basal dendrite (Figs. 23c, 24c). In general, the pyramidal neurons do not reveal any preferential orientation. CR-ir pyramidal cells show less dendritic ramifications than CB-ir pyramidal neurons. CR- and CB-ir pyramids are mainly found in the basal and the lateral nucleus.

Fig. 23a–c. Camera lucida drawings of CB-ir nerve cells within the amygdala of the 8th gestational month. **a** Bipolar nerve cells; **b** multipolar nerve cells; **c** pyramidal nerve cells. Scale bar: 100 μm. (After Setzer and Ulfig 1999, with permission)

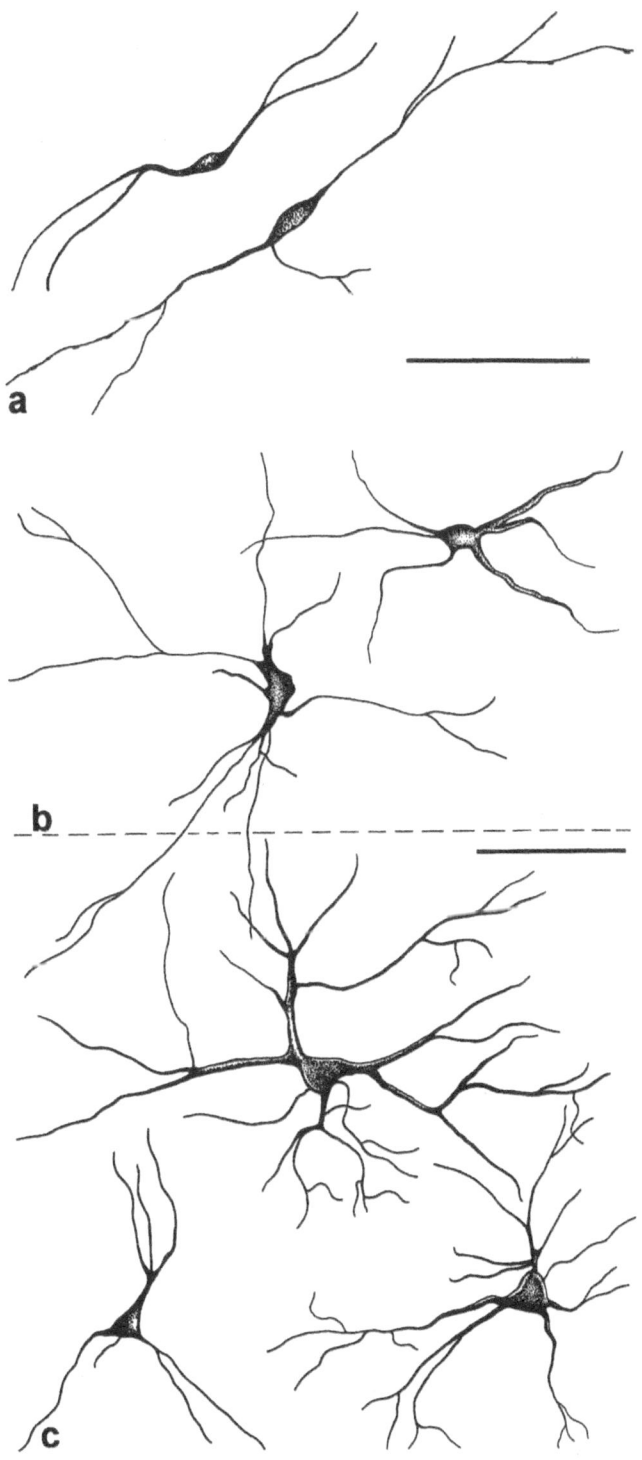

Fig. 24a–c. Camera lucida drawings of CR-ir neurons within the amygdala in the 8th gestational month. a Bipolar nerve cells; b multipolar nerve cells; c pyramidal nerve cells. Scale bar: 100 μm. (After Setzer and Ulfig 1999, with permission)

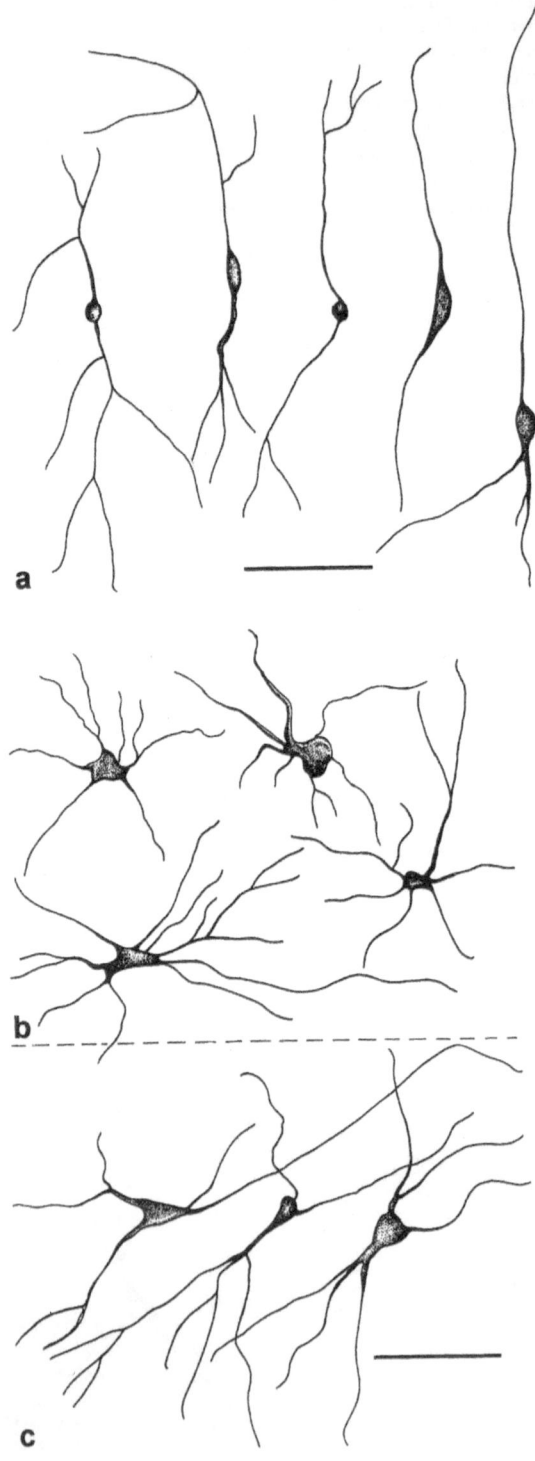

Pyramidal cells representing the projection neurons of the amygdala constitute the predominant nerve cell type in the basolateral amygdala (McDonald 1992). In CB and CR immunopreparations, the number of presumed pyramidal neurons is, however, low to moderate, so only a small percentage of pyramidal cells present in a section is CB- or CR-ir.

One main characteristic of pyramidal cells is their spiny dendritic tree (Braak and Braak 1983). Unfortunately, this feature cannot be judged in fetal specimens. Presumed pyramidal neurons, however, reveal other characteristics typical of pyramidal cells such as shape of the soma, features of dendrites, and the emergence site of the axon.

In the adult human amygdala, pyramidal neurons express CB but not CR (Sorvari et al. 1996a, b). Thus, it is most likely that pyramidal neurons transiently express CR. Similarly, CR has been shown to be transitorily expressed in pyramidal cells of the rat cerebral cortex (Schierle et al. 1997).

7.3.2.3
Nonpyramidal Neurons

The bulk of CB- and CR-ir cells of the human fetal amygdala in the 8th gestational month belong to three classes of nonpyramidal neurons.

1. **Bipolar nerve cells** (Figs. 23a, 24a)
   Bipolar nerve cells are found in even numbers within all amygdaloid nuclei. The dendrites of these spindle-shaped neurons emanate from opposite sides of the cell body and branch at a variable distance from the soma. Most of the bipolar cells are medium-sized, but some show a distinctly small cell body.
2. **Large multipolar neurons** (Figs. 23b, 24b, 25)
   From the polygonal cell body, slender as well as stout dendrites spreading out in all directions emerge. The four to six dendrites show ramifications close to the soma. This nerve cell type is present all over the amygdala.
3. **Small multipolar neurons** (Figs. 23b, 24b, 25)
   This neuronal type is also encountered in all amygdaloid nuclei and is more frequently CR-ir than CB-ir. From the small roundish soma, several dendrites arise, which spread in all directions. Dendritic ramifications are only seldom seen. Only infrequently do the nonpyramidal neurons reveal ir axons or axonal stems.

The number of CR-ir nonpyramidal neurons is distinctly higher than that of CB-ir nerve cells. On the whole, the nonpyramidal CB- and CR-ir nerve cells closely resemble the class II and III neurons described by Braak and Braak (1983) in the adult human brain. Such a correlation between CB- and CR-ir nerve cells in the adult amygdala and those of class II and III neurons has also been demonstrated by Pitkänen and Amaral (1993) and Sorvari et al. (1996a, b). Most probably, the CB- and CR-ir nonpyramidal cells represent local-circuit neurons, corresponding to cortical stellate neurons.

Double-labelings could show only a very rare coexpression of CB and CR in the human fetal amygdala. Thus, CB and CR are mainly in complementary subpopulations of nonpyramidal nerve cells. Similarly, CB and CR are expressed in largely

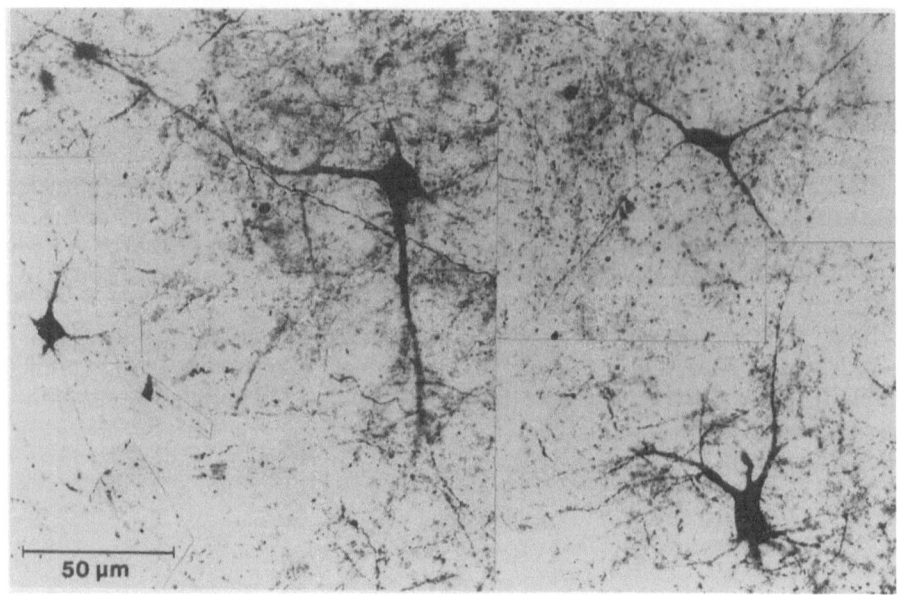

Fig. 25. Microphotographs of small and large multipolar CR-ir neurons (taken from the lateral and basal nucleus of the amygdala), 8th gestational month. (From Setzer and Ulfig 1999, with permission)

complementary subpopulations of stellate interneurons of the hippocampal formation, which express the inhibitory transmitter GABA (Miettinen et al. 1992). Thus, it is apparent that nonpyramidal neurons are neurochemically heterogeneous.

The distinct classes of amygdaloid interneurons are likely to be functionally different because they most probably have different input-output characteristics. Based on detailed data on neuronal circuitry in the hippocampal formation (Jiang and Swann 1997), the following way of controlling pyramidal cells in the amygdala has been proposed (Sorvari et al. 1998): CR-ir interneurons form inhibitory synapses on CB-ir interneurons which, in turn, make inhibitory contacts on the pyramidal cell. Thus, CR-ir interneurons may influence pyramidal cell activity via a disinhibitory interneuronal pathway.

When comparing the CR immunosections of the 5th gestational month with those of the 8th month, it is obvious that the basolateral nuclei of the 5th month contain a distinctly higher number of CR-ir nerve cells than the corticomedial nuclei. In the 8th month, however, the corticomedial nuclei also contain a high number of CR-ir neurons. This finding may be related to the process of migration. In the 5th month, neurons migrating through the basolateral nuclei have not yet reached the corticomedial nuclei.

7.4
Diffuse Calbindin and Calretinin Immunostaining in the Human Fetal Amygdala

Diffuse (punctate or neuropil) immunolabeling in the amygdala is nuclear-specific and is likely to reflect differential afferent input of the various amygdaloid nuclei and areas (Setzer and Ulfig 1999).

7.4.1
Diffuse Labeling in the 5th–6th Gestational Month

The intensity of diffuse CB and CR immunolabeling distinctly varies between the amygdaloid nuclei and areas.

In CB immunosections, the highest density of diffuse immunostaining is present in the central nucleus. The basolateral nuclei as a whole exhibit a moderate density of ir puncta. The intercalate nuclei also show a moderate intensity of diffuse immunoreactivity. In the remaining subdivisions of the amygdala, only weak immunostaining is observed (Figs. 16, 17).

In CR immunopreparations, the central and medial nucleus as well as the anterior and posterior cortical nucleus reveal a conspicuously high density of ir puncta. A moderate density of puncta is observed in the lateral nucleus, the anterior amygdaloid area, and the amygdalohippocampal area. The remaining subdivisions of the amygdala show low diffuse CR immunostaining (Fig. 18).

7.4.2
Diffuse Labeling in the 8th–9th Gestational Month

Distinct differences can be observed when comparing diffuse CB and CR labeling in the 5th–6th and 8th–9th months.

In CB immunopreparations, intense punctate immunolabeling is found in the medial nucleus, the cortical nucleus, and the periamygdaloid cortex. The central nucleus exhibits a moderate number of puncta. Medium to low levels of diffuse labeling are observed in the remaining nuclei and areas (Fig. 21).

In CR immunosections of the 8th–9th gestational month, a high amount of ir puncta is detected in the anterior cortical nucleus, the medial nucleus, and the amygdalohippocampal area. Moderate punctate labeling is present in the lateral, central, paralaminar, and posterior cortical nuclei. The remaining nuclei and areas show weak diffuse labeling (Fig. 22).

In general, intense diffuse CR and CB immunostaining is observed in nuclei and areas which only exhibit a low number of immunostained nerve cells. Thus, CR and CB may be contained in projection neurons of other brain areas that provide afferent input to the amygdala. Various subcortical nuclei projecting to the amygdala contain CR or CB.

The periaqueductal gray of the mesencephalon containing a high number of CB-ir neurons is reciprocally connected with the central nucleus (De Leon et al. 1994; Rizvi et al. 1991). Accordingly, the central nucleus is characterized by intense punctate CB immunolabeling in the 8th–9th gestational month (see preceding paragraphs in this section).

The cholinergic magnocellular nuclei of the basal forebrain, which contain a high number of CB-ir neurons, project mainly to the basal nucleus (Celio 1990; Amaral et al. 1992). The latter displays a large amount of CB-ir puncta already in the 5th–6th gestational month. This observation is in line with the fact that the magnocellular nuclei of the basal forebrain provide early afferent input to the amygdala (Kostovic 1986).

The hypothalamic tuberomamillary nucleus, which projects to various amygdaloid nuclei, also reveals CR-ir neurons (Panula et al. 1989; Borhegyi and Leranth 1997; Kiss et al. 1997).

When comparing the immunosections of the 5th–6th and 8th–9th gestational months, distinct differences in the distribution patterns of diffuse immunoreactivity are evident. Thus, the mature pattern of diffuse immunolabeling is reached only after a prolonged period of development which is characterized by increases as well as decreases in diffuse immunostaining in the various amygdaloid nuclei and areas. An increase in diffuse labeling between the 5th–6th and 8th–9th gestational months could be related to the sequential arrival of afferents, for instance from the basal forebrain nuclei first and, later on, from the tuberomamillary nucleus. The two nuclear complexes are known to differentiate at different stages of development (Paul and Ulfig 1998).

A decrease in punctate immunolabeling may be due to fiber-diluting effects, such as differentiation of dendritic trees, occurrence and maturation of glial cells, and myelin sheaths or arrival of other afferents that are not CR- or CB-ir. Of course, a decrease in puncta may also indicate a transient CR- or CB-ir projection to the amygdala.

When comparing the findings of diffuse CR- and CB-immunostaining in the 8th–9th gestational month with those in the adult human brain (Sorvari et al. 1996a), only minor differences are evident. Thus, the amygdala has reached a high degree of maturity in the 8th gestational month. Therefore, it may be assumed that only subtle reorganization of, for instance, afferent projections are to be expected during proceeding development.

8 Basal Nucleus of Meynert

8.1
Description

Enlargement of the telencephalic cortex during phylogenesis is associated with an analogous development of subcortical nuclei, which provide extrathalamic afferents to the cerebral cortex. One of these nuclei belonging to the magnocellular complex of the basal forebrain is the basal nucleus of Meynert, which is embedded in the substantia innominata (Ulfig 1989). This nucleus provides the major source of the cholinergic afferents to virtually all areas of the cerebral cortex.

The early differentiation of the basal nucleus and its projections to the cerebral cortex has been demonstrated in the human developing brain using acetylcholinesterase histochemistry (Kostovic 1986). Already at the beginning of the 3rd gestational month, strong acetylcholinesterase activity is seen in the basal nucleus. This observation is in accordance with early cytoarchitectonic investigations in man (Kodoma 1929). In the 5th gestational month, acetylcholinesterase-positive fibers are found throughout the future white matter of the cerebral hemisphere. Strongly reactive fibers can be followed from the basal nucleus, through the external capsule towards the subplate of the frontal, temporal, parietal, and occipital lobes.

8.2
Calbindin Immunoreactive Neurons
in the Basal Nucleus of Meynert

Anti-CB has been shown to be a convenient marker of basal nucleus neurons in the monkey and the human adult (Celio and Norman 1985). In CB immunopreparations of the 7th gestational month, the basal nucleus of Meynert stands out clearly on account of its intense immunostaining. The basal nucleus is composed of two subnuclei displaying a high packing density of ir neurons (Fig. 26). The anteromedial subnucleus, which shows a band-like shape, is mainly located beneath the globus pallidus (Fig. 27). In more posterior sections, the anteromedial subnucleus gradually diminishes and the posterolateral subnucleus gains in size. It occupies a triangular or semilunar field close to the medial margin of the anterior commissure. Mainly, it is located below the putamen and above the amygdala. The subnuclei are surrounded by scattered nerve cells forming the substantia innominata. Scattered neurons are also observed within the medullary laminae between the lateral and medial part of the

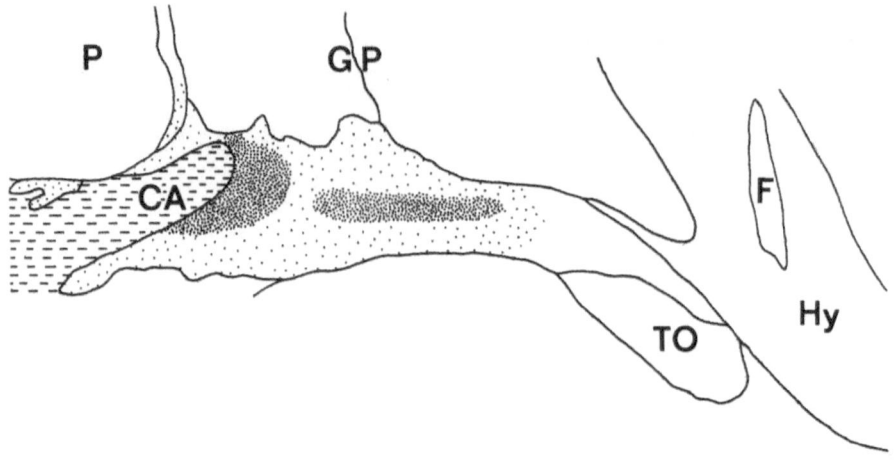

Fig. 26. Two subnuclei of the basal nucleus of Meynert (anteromedial and posterolateral subnuclei, *densely dotted*) embedded in the substantia innominata (*lightly dotted*) and the surrounding brain areas. *CA*, anterior commissure; *F*, fornix; *GP*, globus pallidus; *Hy*, hypothalamus; *P*, putamen; *TO*, optic tract

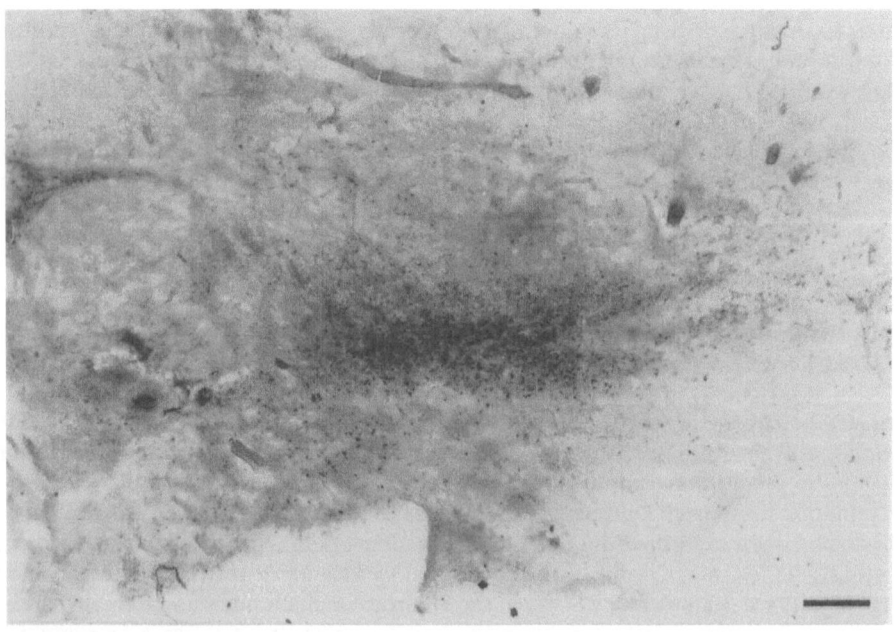

Fig. 27. CB immunopreparation showing the anteromedial subnucleus of the basal nucleus of Meynert, 7 weeks of gestation. Scale bar: 100 μm

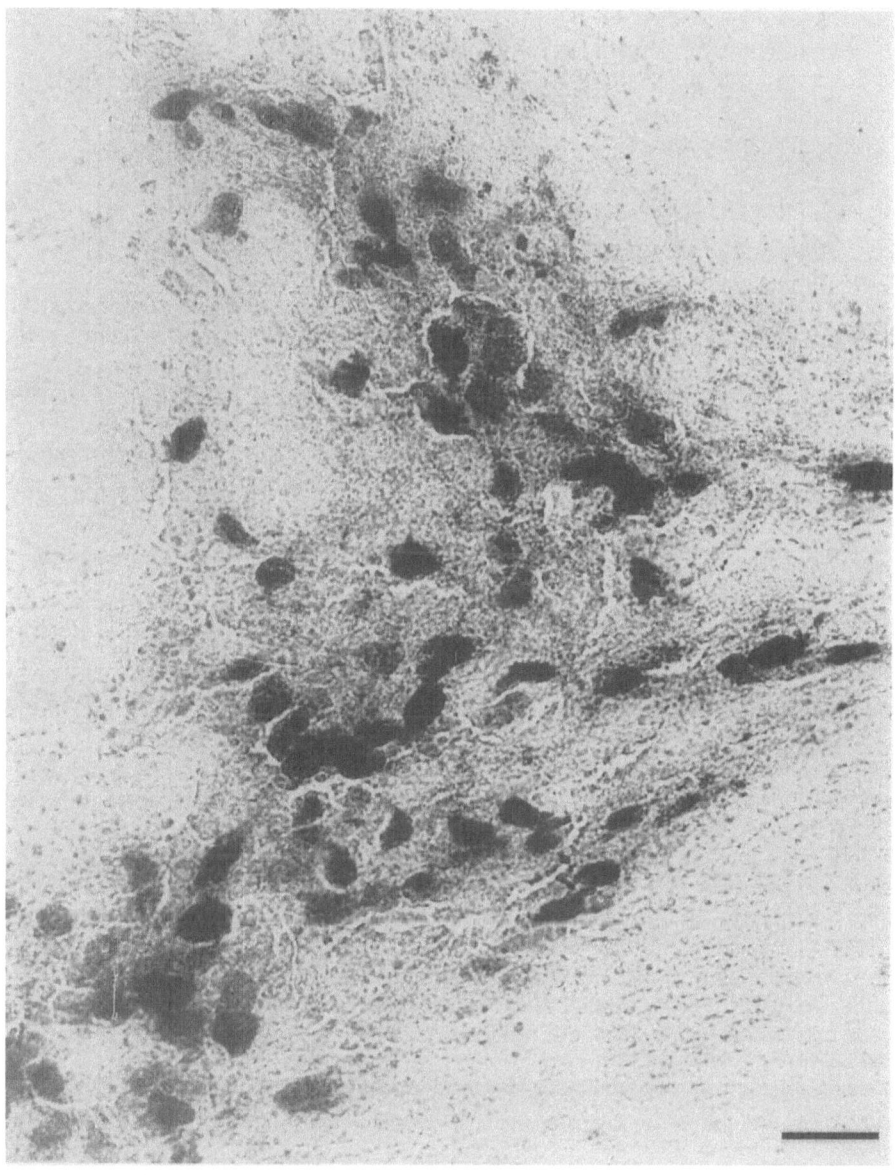

Fig. 28. CB-ir neurons of the basal nucleus (Meynert) next to the anterior commissure, 7 weeks of gestation. Scale bar: 50 μm

globus pallidus as well as between the putamen and the lateral part of the globus pallidus.

Both subnuclei of the basal nucleus and the surrounding substantia innominata are mainly composed of multipolar nerve cells (Fig. 28); a moderate number of bipolar nerve cells is also seen. Immunostaining of the neurons is outstandingly intense.

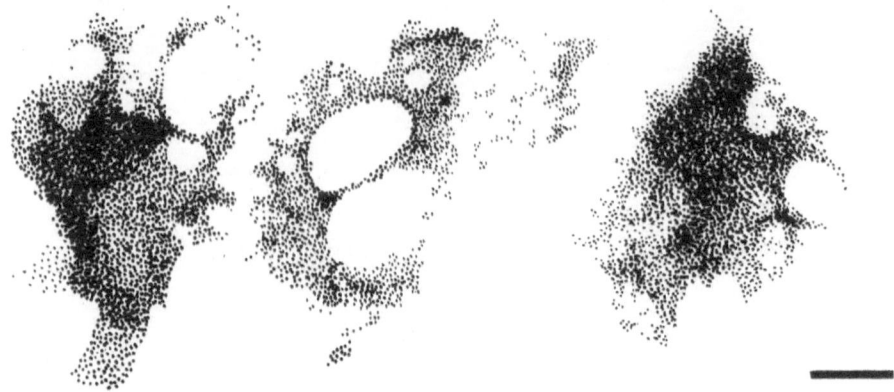

Fig. 29. Drawing of neurons of the basal nucleus revealing vacuolation as seen in CB immunopreparations of hydrocephalic brains. Scale bar: 10 µm

8.3
Effect of Fetal Hydrocephalus on Neuronal Morphology in the Basal Nucleus of Meynert

In experimental hydrocephalus, a progressive functional injury of the cholinergic system has been described (Tashiro and Drake 1998). To determine the effect of fetal hydrocephalus on basal nucleus neurons in the developing human brain, CB immunopreparations of five hydrocephalic cases ranging in age from 22 to 34 weeks of gestation were investigated.

The basal nucleus in hydrocephalic brains harbors a moderate number of vacuolated neurons (Fig. 29). The latter are randomly distributed within the two subnuclei and, occasionally, in the substantia innominata. Vacuolated neurons are never found in age-matched controls (Ulfig 2002).

Vacuolation within neurons has been demonstrated to occur as a degenerative sign after fiber tract degeneration or after axotomy (Kreutzberg et al. 1997). Vacuoles may arise from smooth endoplasmatic reticulum or the Golgi apparatus. Thus, they may be suggestive of perturbations in synaptic protein trafficking.

Vacuolation in the basal nucleus as seen in fetal hydrocephalic brains is likely to be caused by damage to the axons of the basal nucleus neurons within the intermediate zone.

The intellectual impairment observed in patients despite shunt insertion may partly be related to pathological alterations in the basal nucleus (Raimondi and Soare 1974). Accordingly, evidence has been provided that irreversible damage of the cholinergic system occurs in long-lasting experimental hydrocephalus (Miyajima et al. 1996).

9 Hypothalamic Tuberomamillary Nucleus

9.1
Description

The hypothalamic nuclei occupy inferior portions of the diencephalon below the hypothalamic sulcus. The hypothalamus extends from the lamina terminalis to a vertical plane posterior to the mamillary bodies, and it can be subdivided into three major (frontal) regions (in an anteroposterior sequence): the chiasmatic region anteriorly followed by the tuberal region and the mamillary region posteriorly.

The magnocellular tuberomamillary nucleus extends through the tuberal and mamillary region (Fig. 30). In the tuberal region, the tuberomamillary nucleus nearly

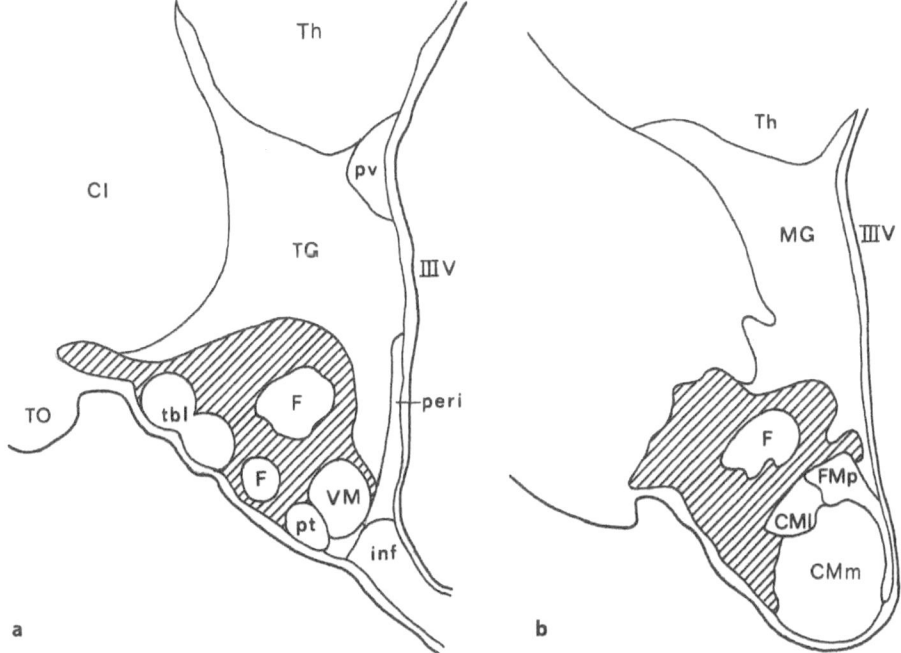

Fig. 30 a, b. The tuberal **a** and mamillary region **b** of the adult hypothalamus. The tuberomamillary nucleus is *hatched*. *C*, Internal capsule; *CMl*, mamillary body, lateral nucleus; *CMm*, mamillary body, medial nucleus; *F*, fornix; *FMp*, principal mamillary fascicle; *inf*, infundibular nucleus; *MG*, mamillary gray; *peri*, periventricular nucleus; *pv*, paraventricular nucleus; *pt*, paratuberal nucleus; *Th*, thalamus; *TG*, tuberal gray; *TO*, optic tract; *IIIV*, third ventricle; *VM*, ventromedial nucleus

completely surrounds the lateral tuberal nucleus, lies around the fornix (perifornical portion) and, in part, around the ventromedial nucleus. Laterally, it extends upon the optic tract. In the mamillary region, the tuberomamillary nucleus is located laterally and superiorly to the mamillary nuclei (the paramamillary and supramamillary portion).

The tuberomamillary nucleus is characterized by distinctly large neurons; their packing density varies considerably. Central portions of the nucleus display densely packed nerve cells, whereas the periphery generally shows lower packing densities of neurons. On account of this low packing density, the tuberomamillary nucleus appears ill defined.

Climbing up the phylogenetic scale, the tuberomamillary nucleus becomes more extensive relative to the other hypothalamic nuclei, and it is most extensive in man. The tuberomamillary nucleus shows many similarities with the magnocellular nuclei of the basal forebrain (basal nucleus of Meynert). Both nuclear complexes are composed of outstandingly large nerve cells, both complexes are characterized by cell-dense areas, which are embedded in cell-sparse portions, and both complexes are most extensively developed in man. Furthermore, the two complexes have widespread projections and provide the major source of nonthalamic projections to the cerebral cortex (Saper 1990; Panula and Airaksinen 1991; Köhler et al. 1985).

The tuberomamillary nucleus in the adult is the only brain region revealing histamine-ir nerve cells. The tuberomamillary neurons contain a variety of additional neuroactive substances such as GABA, galanin, adenosin (Panula et al. 1989). In the human developing and mature brain, reliable demonstration of the aforementioned substances is rendered difficult. To investigate the ontogenesis of the tuberomamillary nucleus, it is advantageous to have a convenient marker.

9.2
Parvalbumin and Calretinin in the Tuberomamillary Nucleus

Anti-PV has been shown to immunolabel the adult tuberomamillary nucleus intensely and completely. In such immunopreparations, the nucleus can be clearly delineated. At high magnification, the large neurons appear multipolar; the polygonal cell body generates slender dendrites that spread out in all directions (Fig. 31). On average, four to ten dendrites emerge; ramifications are seen close to the soma. PV-ir neurons are found in all portions of the tuberal and mamillary part of the tuberomamillary nucleus. The nucleus stands out clearly because adjacent nuclei are mostly devoid of PV-ir structures (Ulfig et al. 1989).

The histaminergic innervation of the cerebral cortex has so far not been considered adequately in developmental studies. It is not known when the afferents from the tuberomamillary nucleus reach the subplate.

During a prolonged period of development, cholinergic fibers from the basal nucleus of Meynert and thalamic afferents intermingle within the subplate (Kostovic 1986). It has to be determined whether there is an overlapping of thalamic and histaminergic input in the subplate and what the functional role can be ascribed to such intermingling of different fiber systems within the subplate.

Moreover, detailed data on the maturation of tuberomamillary neurons are so far not available. Preliminary results suggest that differentiation of tuberomamillary

Fig. 31. Camera lucida drawings of PV-ir neurons of the tuberomamillary nucleus in the adult brain. Scale bar: 50 μm

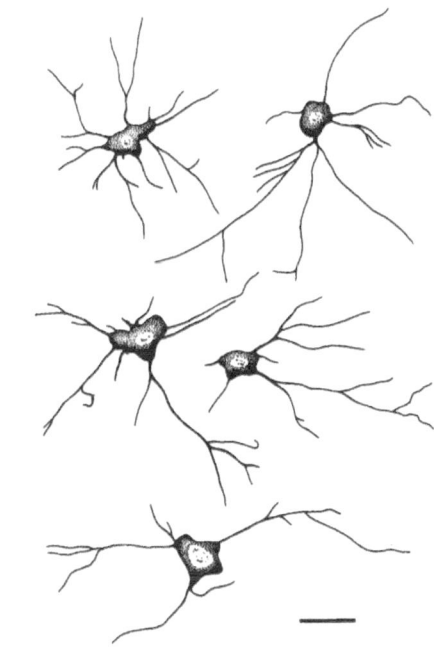

Fig. 32. Camera lucida rawings of PV-ir neurons of the tuberomamillary nucleus, 7th gestational month. Scale bar: 50 μm

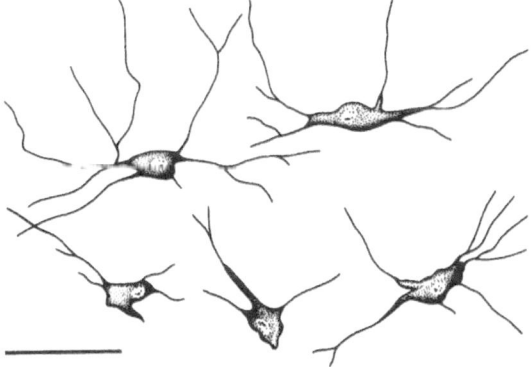

neurons occurs distinctly later than that of the functionally related basal nucleus of Meynert. Maturely appearing PV-ir neurons do not appear in the tuberomamillary nucleus before the 7th gestational month. The PV-ir neurons mainly display a multipolar shape; dendrites generally bifurcate close to the soma (Fig. 32). At the end of the 7th gestational month, PV-ir neurons appear densely packed within the entire territory of the tuberomamillary nucleus.

In addition to PV, neurons of the supramamillary portion of the tuberomamillary nucleus have been shown to express CR (Kiss et al. 1997; Borhegyi and Leranth 1997). In the 8th gestational month, CR-ir nerve cells are seen at the superior margin of the

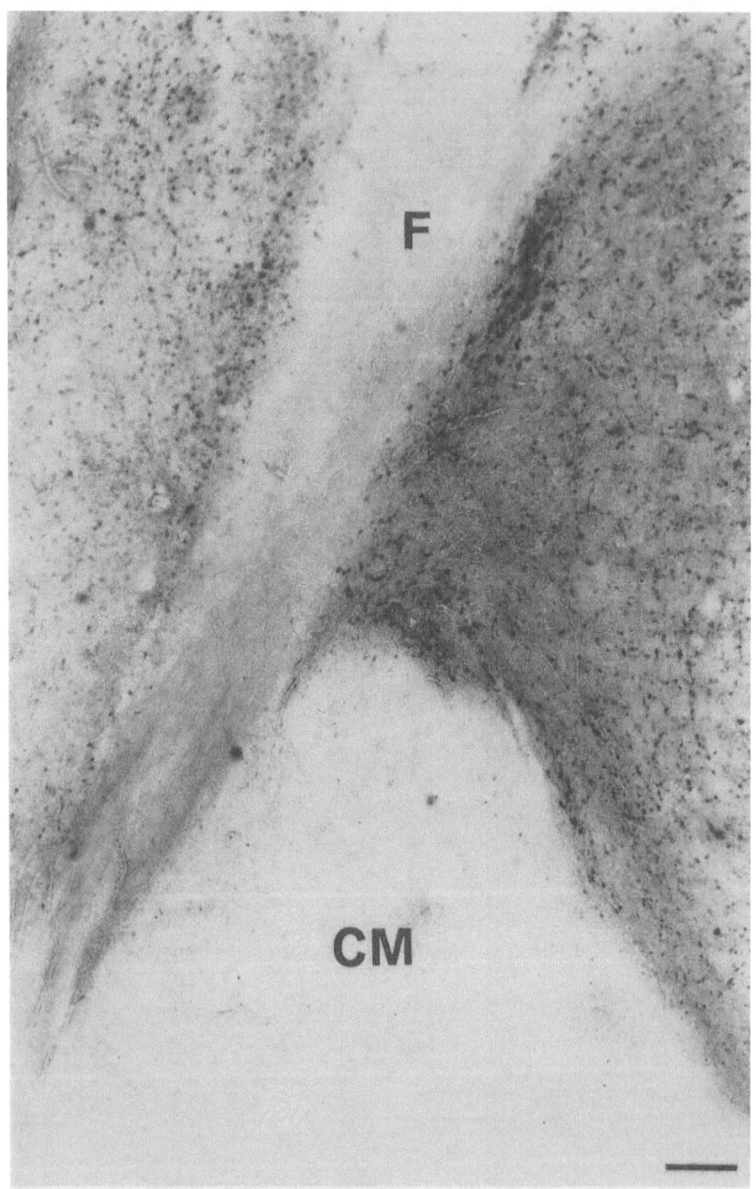

Fig. 33. Supramamillary portion of the tuberomamillary nucleus as seen in CR immunopreparations. *CM*, mamillary body; *F*, fornix. Scale bar: 200 μm

mamillary body; these neurons are densely packed and surround the fornix (Fig. 33). These CR-ir neurons, which more probably coexpress PV, have been demonstrated to form supramamillohippocampal projections. Thus, tuberomamillary neurons projecting to the hippocampus differ from those projecting to isocortical areas with regard to their expression of CaBPs.

10 Thalamic Reticular Complex

10.1
Description

The thalamic reticular nucleus is a sheet of neurons which enfolds the dorsal thalamus laterally. It lies between the external medullary lamina and the internal capsule. The superior and inferior poles of the nucleus bend medially; thus, the reticular nucleus gets its shell-like shape. The thalamocortical and corticothalamic fibers, which all have to traverse the reticular nucleus, give collaterals to this nucleus (Jones 1985). Moreover, afferents of the reticular nucleus come from the cholinergic nuclei of the basal forebrain and from various brain stem nuclei (Cornwall et al. 1990; Parè et al. 1990). The reticular nucleus is composed of GABAergic neurons which provide inhibitory projections to the dorsal thalamic nuclei.

In the adult human brain, the reticular nucleus appears inconspicuous because it contains only a low number of nerve cells. In the fetal human brain, however, the reticular nucleus is characterized by a high packing density of nerve cells and therefore appears as a prominent structure. Another outstanding feature in the developing brain is the presence of neurons that closely resemble reticular neurons within the internal capsule. This group of neurons, referred to as the periventricular nucleus, is found between the fibers of the internal capsule lateral to the reticular nucleus and medial to the globus pallidus and/or the putamen. This perireticular nucleus has been established as a distinct entity during development (Mitrofanis and Guillery 1993).

The human fetal reticular complex (Fig. 34) is composed of four architectonic subdivisions: the main portion, the perireticular nucleus, the medial subnucleus, and the pregeniculate nucleus (Ulfig et al. 1998b).

10.2
Main Portion

In PV immunopreparations from fetal brains of the 6th and 7th gestational months, the main portion (of the reticular complex) appears as a conspicuously broad gray area displaying a distinctly high packing density of PV-ir nerve cells (Fig. 35). Moreover, PV-ir fibers coming from the main portion are seen to enter the dorsal thalamus. This projection from the main portion is among the earliest afferents that the dorsal thalamus receives (Mitrofanis 1994). Furthermore, the early fibers from the main portion may represent pioneer axons, which could be responsible for the guidance of outgrowing thalamic axons or of cortical fibers entering the dorsal thalamic nuclei.

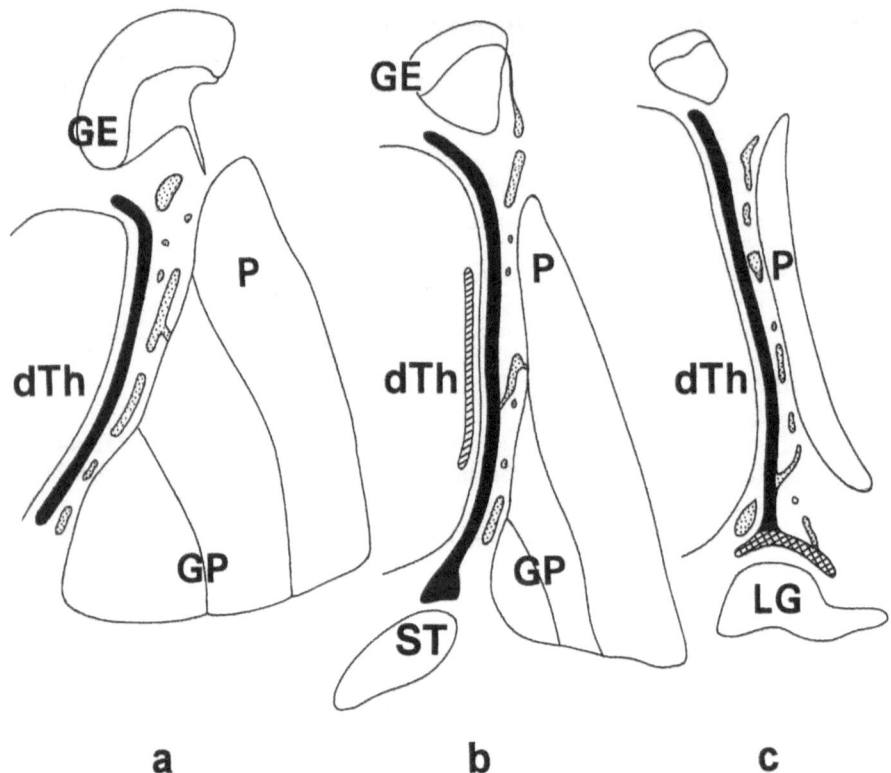

Fig. 34 a–c. Frontal sections of the **a** anterior, **b** middle, and **c** posterior third of the thalamus, 7th gestational month. The four architectonic subdivisions of the thalamic reticular complex are marked: *black*, main portion; *dotted*, perireticular nucleus; *hatched*, medial subnucleus; *double hatched*, perigeniculate nucleus. *dTh*, dorsal thalamus; *GE*, ganglionic eminence; *GP*, globus pallidus; *LG*, lateral geniculate body; *P*, putamen; *ST*, subthalamic nucleus. (After Ulfig et al. 1998b, with permission)

Comparing the fetal with the adult main portion, it is obvious that it is surprisingly large during development. With increasing maturity, it is distinctly reduced in size, probably as a consequence of nerve cell death. Therefore, it can be assumed that it fulfils important developmental functions.

10.3
Perireticular Nucleus

In PV and CR immunopreparations of the 6th and 7th gestational months, a considerable number of ir neurons are visible within the internal capsule. Perireticular neurons lie singly or form differently shaped clusters consisting of variable numbers of nerve cells (Figs. 35, 36).

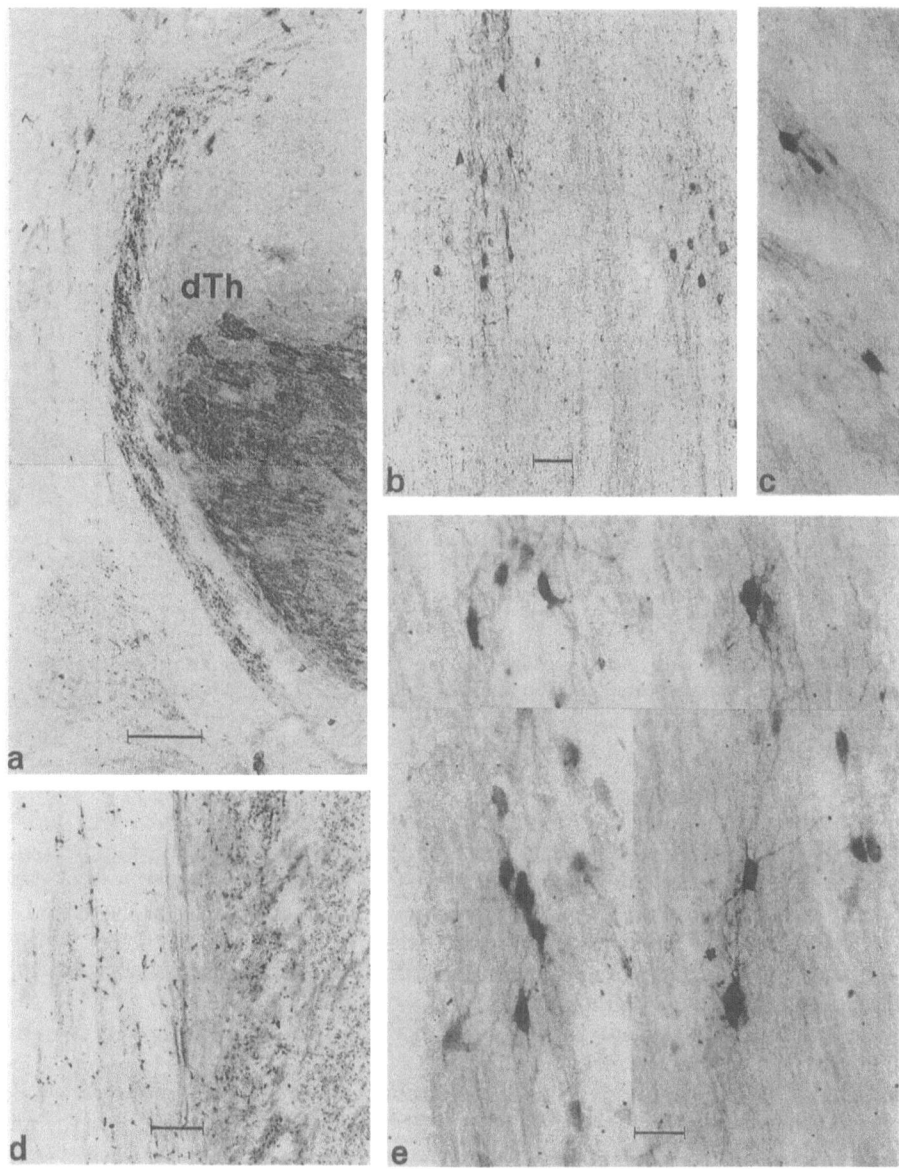

Fig. 35 a–e. PV-ir neurons in the main portion and the perireticular nucleus of the thalamic reticular complex, 6th gestational month. **a** The main portion revealing a high packing density of ir neurons form a broad shell surrounding the lateral margin of the dorsal thalamus (*dTh*). **b** Two groups of perireticular neurons. **c** Perireticular neurons. **d** Perireticular neurons (*left*) are in continuity with the main portion (*right*). **e** Perireticular neurons at high magnification. Scale bars: **a**, 1 mm; **b**, and **d**, 200 μm; **e** and **c**, 50 μm. (From Ulfig et al. 1998b, with permission)

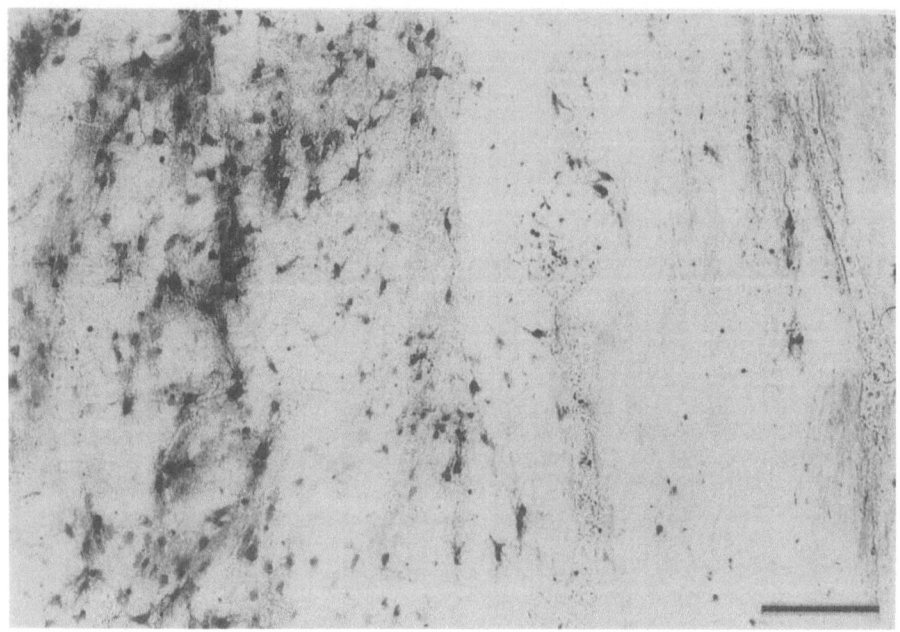

Fig. 36. CR-ir neurons in the main portion (*left*) are in continuity with the perireticular nucleus (*right*) of thalamic reticular complex, 8th gestational month. Scale bar: 400 µm

Often the perireticular neurons are separated from the main portion by a nerve cell-free area, sometimes the perireticular nucleus merges with the main portion. Merging of the perireticular nucleus with the globus pallidus and the medial and lateral medullary lamina (between the two segments of the globus pallidus and between the globus pallidus and the putamen, respectively) is also seen (Fig. 35). The superior pole of the perireticular nucleus is observed to directly border the margin of the ganglionic eminence (Ulfig et al. 1998b).

The perireticular nucleus is found along the entire anteroposterior as well as superioinferior axis of the main portion (Fig. 34).

Evidence has been provided that the perireticular neurons are among the earliest neurons generated in the thalamus (Earle and Mitrofanis 1996). Mostly, both the main portion and the perireticular nucleus stem from the diencephalic proliferative area lining the third ventricle. It has been postulated that at least some perireticular neurons derive from telencephalic proliferative regions. In line with this notion, perireticular neurons are found in very close vicinity to the telencephalic ganglionic eminence (see preceding paragraphs).

In the rat brain, a dramatic decline of perireticular neurons has been shown to occur during a postnatal 10-day period (P5–P15) and to lead to a disappearance of 98% of the perireticular neurons. A dilution effect due to growth of white matter is likely to contribute to this decrease. However, the enormous cell loss cannot solely be caused by an dilution effect. Two other mechanisms underlying the sharp decline in neuronal density have been evidenced: neuronal death and migration of perireticular neurons into the globus pallidus (Earle and Mitrofanis 1996). In the developing globus pallidus, neurons are discernible that greatly resemble perireticular neurons with regard to their morphology and their expression of PV. Later these neurons can no longer be detected; most probably they undergo cell death.

In the human brain, perireticular neurons also disappear during postnatal development. Only very few perireticular neurons have been shown in a 1-year-old infant (Letinic and Kostovic 1996a). Thus the perireticular nucleus is to be regarded as a transient structure of the developing human brain.

Various studies have provided evidence that the perireticular nucleus is involved in guiding axons (Mitrofanis and Guillery 1993; Ramcharan and Guillery 1997; Ulfig et al. 2000c). The perireticular nucleus most probably serves as an intermediate target for growing axons which are here sorted and directed towards their different final targets. Outgrowing cortical axons, which reach the perireticular nucleus, are either destined to grow towards the dorsal thalamus or towards the brain stem. In the animal brain, the course of developing cortical axons has been demonstrated using DiI (carbocyanine dye). A number of fibers sharply turn towards the dorsal thalamus as they reach the perireticular nucleus.

Moreover, these fibers display an interweavement within the perireticular nucleus. Another group of fibers directly runs toward the cerebral peduncle without changing course or revealing interweavement in the perireticular nucleus. These axons belong to the corticobulbar or corticospinal tract. A similar arrangement of axons has also been demonstrated in the human fetal brain using the axonal marker SMI 312 (Ulfig and Chan 2002b). In SMI 312/CR double-labelings, CR-ir perireticular neurons are surrounded by dense fiber accumulations. The latter, corresponding to the above-mentioned interweavement of fibers, does not cover the somata of neurons, but are in apposition to the dendritic trees of perireticular neurons. These observations are in accordance with previous data showing that corticothalamic fibers make contact upon dendrites of the perireticular nucleus (Ulfig et al. 2000d). With proceeding development, the axonal pattern changes significantly. From the 8th gestational month onwards, an interweavement of axons is no longer detectable. On the whole, the data derived from studies on animal as well as human brains show that within the perireticular nucleus a sorting of cortical axons takes place.

10.4
Medial Subnucleus

The medial subnucleus is present in frontal sections through the middle third of the fetal thalamus. It is a band-like structure between the main portion and the dorsal thalamus; thin cell-sparse stripes separate the medial subnucleus from these adjacent structures.

Double-immunolabelings with anti-PV and anti-CR clearly demonstrate the medial subnucleus (Fig. 37a). In these immunopreparations, the medial subnucleus stands out prominently because its neurons express only CR. Thus, it is chemoarchitectonically different from the main portion. It does not express GABA or PV (Clemence and Mitrofanis 1992). The medial subnucleus may well correspond to the inner small-cell region described by Clemence and Mitrofanis (1992) because the somata of its neurons appear smaller than those in the main portion (Ulfig et al.

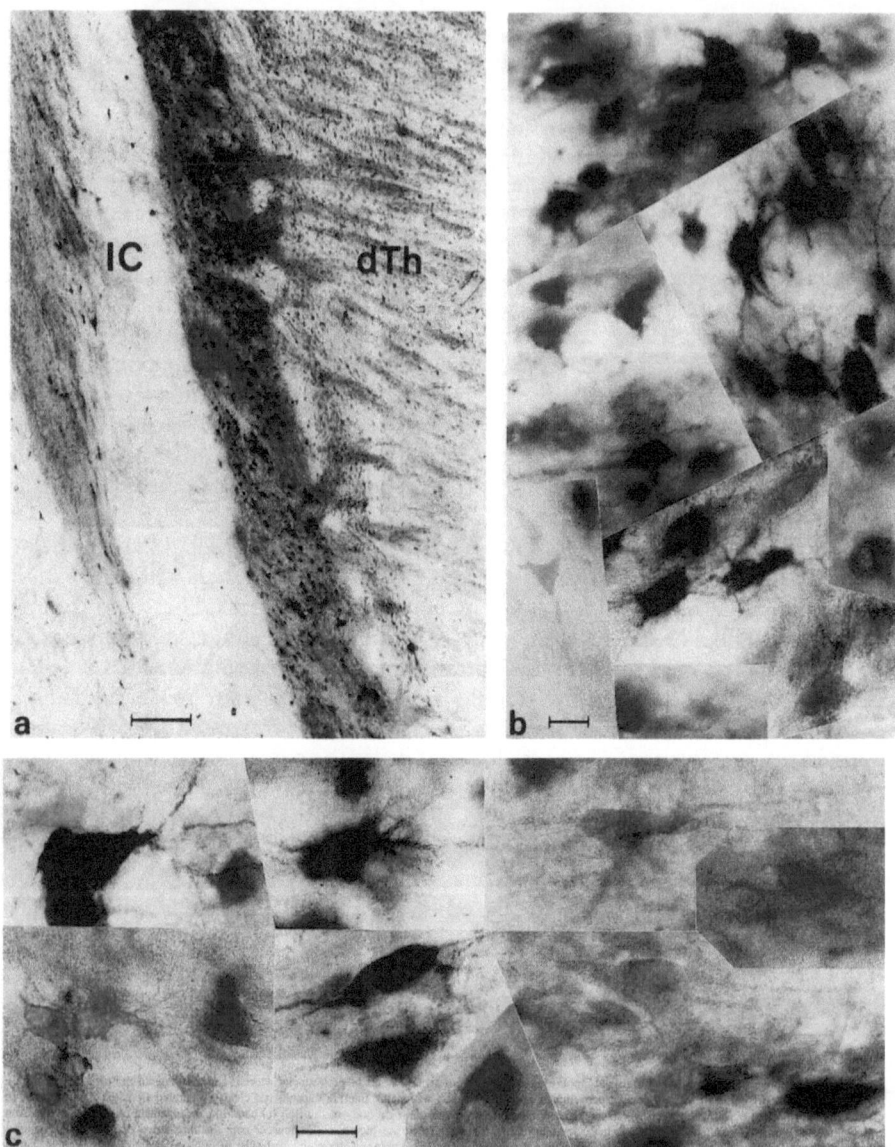

Fig. 37 a–c. CR + PV double-labeling of the thalamic reticular complex, 6th gestational month. **a** Survey magnification: main portion (*black*, double-labeled, CR + PV-ir) and medial subnucleus (*brown*, single-labeled, CR-ir). Note the *blue* fiber bundles within the dorsal thalamus. *IC*, internal capsule; *dTh*, dorsal thalamus. **b, c** Neurons from the main portion and the perireticular nucleus: double-labeled neurons are *black*, solely PV-ir neurons *blue*, and solely CR-ir neurons *brown*. Scale bars: **a**, 200 μm, **b** and **c**, 20 μm. (From Ulfig et al. 1998b, with permission)

1998b). It is so far not clear whether this medial subnucleus represents an extension of the zona incerta. Possibly, the medial subnucleus provides a projection to the centromedian and parafascicular nuclei of the dorsal thalamus (Clemence and Mitrofanis 1992; Royce et al. 1991).

10.5
Pregeniculate Nucleus

The pregeniculate nucleus is a cap-like structure located above the superior margin of the lateral geniculate body (Fig. 34). The neurons of the pregeniculate nucleus are arranged in two band-like structures (inferior and superior lamina) lying parallel to the superior margin of the lateral geniculate body. The superior lamina is continuous with the main portion and the perireticular nucleus.

This architectonic organization can be observed in PV immunopreparations of the 6th and 7th gestational months.

This architectonic component of the thalamic reticular complex may also act as a transient target region of growing axons. It may receive afferents from the retina, which are later relocated to the geniculate body (Linden et al. 1981).

10.6
Neuronal Types of the Thalamic Reticular Complex as Seen in Parvalbumin and Calretinin Immunopreparations

Various nerve cell types are present in all subdivisions of the reticular complex: bipolar, triangular as well as multipolar neurons are encountered. The soma size varies to a considerable extent. No correlation between morphology (bipolar, triangular, multipolar) or soma size and the expression of CR and PV can be detected (Ulfig et al. 1998b). Within the main portion, the long dendrites are mainly oriented parallel to the lateral surface of the dorsal thalamus (Figs. 38–40).

Furthermore, an outstanding neuronal type is seen. It is characterized by a huge number of thin dendrites which arborize extensively at short distances from the soma (bushy dendritic trees). These bushy dendritic trees often arise from only one side of the cell body (Fig. 38). This nerve cell type is encountered in the main portion and in the perireticular nucleus. The elaborated dendritic trees of this neuronal type could provide a substratum for extensive interactions with growing axons. The profuse dendritic trees could also be regarded as a degenerative sign. A similar morphological appearance has been described for Cajal-Retzius cells of the cortical molecular layer (Meyer and Gonzalez-Hernandez 1993). A large number of Cajal-Retzius cells and reticular neurons are transient in nature.

In immunopreparations double-labeled with anti-CR and anti-PV, it is obvious that the majority of neurons in the main portion as well as in the perireticular nucleus expresses both CR and PV (Fig. 37). Intermingled between these double-labeled nerve cells, a few neurons are found that solely express either CR or PV. A similar immunolabeling pattern can be observed in the pregeniculate nucleus.

Fig. 38 a, b. Camera lucida drawings of **a** PV-ir and **b** CR-ir neuronal types within the main portion of the thalamic reticular complex, 7th gestational month. A neuron characterized by a bushy dendritic tree is marked by a *star*. Scale bar: 20 μm

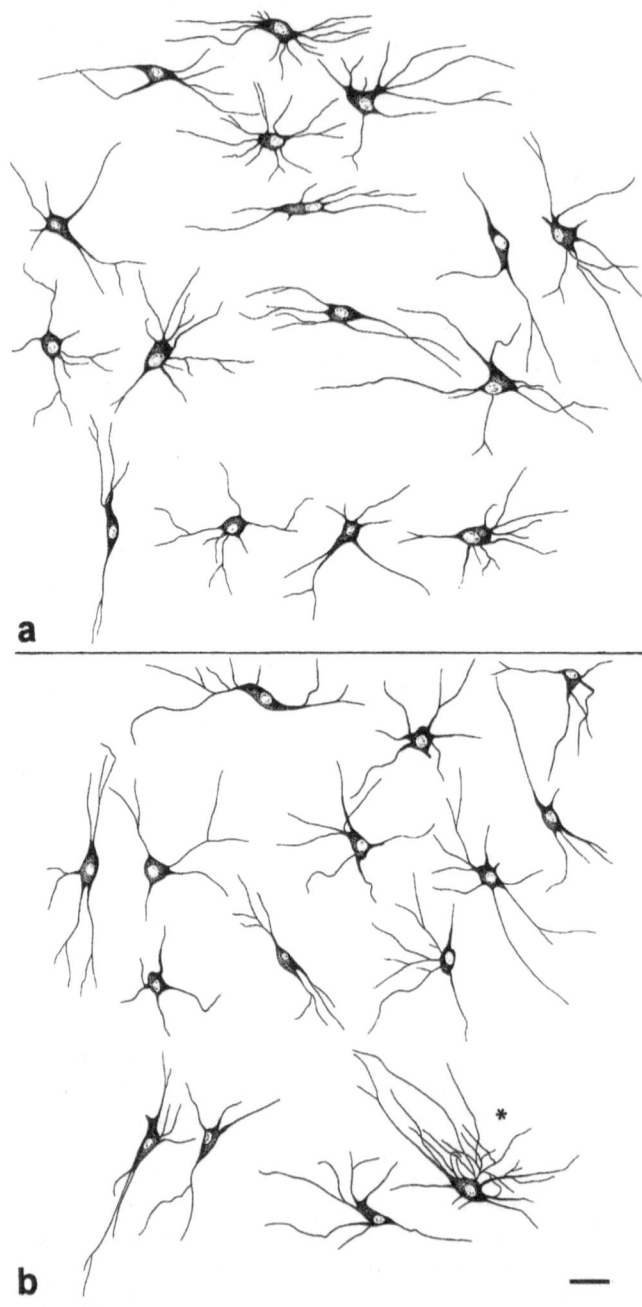

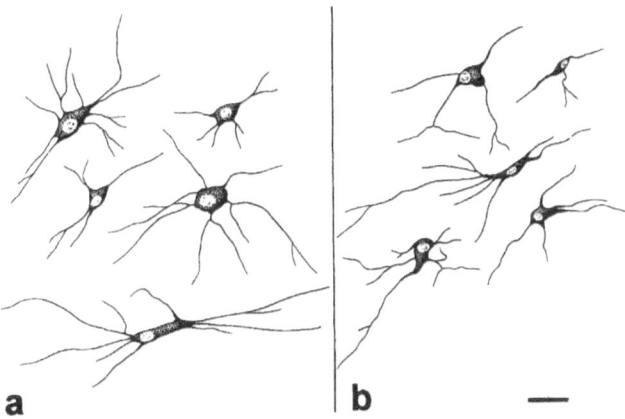

Fig. 39 a, b. Camera lucida drawings of **a** PV-ir and **b** CR-ir nerve cell types within perireticular nucleus, 7th gestational month. Scale bar: 20 μm

a   b

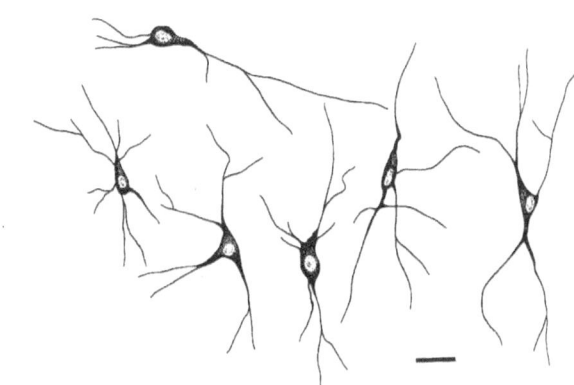

Fig. 40. Camera lucida drawings of CR-ir nerve cells observed in the medial subnucleus of the thalamic reticular complex, 7th gestational month. Scale bar: 20 μm

The medial subnucleus, however, contains only CR-ir nerve cells, mostly revealing a small soma size (Fig. 40). This subdivision is completely devoid of PV-ir nerve cells and, thus, differs from the other three subdivisions of the reticular complex.

The main portion and the perireticular nucleus has been shown to express the same neuroactive substances in the rat. So the coexpression of CR and PV in most of their neurons further underlines the notion that perireticular neurons as a whole appear as outriders of the main portion.

The observation that the bulb of neurons in the main portion coexpresses PV and CR is surprising because in the adult brain CR-ir neurons only represent a subpopulation of main portion neurons and are restricted to certain parts of the main portion (Lizier et al. 1997). Therefore, a transient expression of CR in this nucleus becomes obvious.

11 Red Nucleus

11.1
Description

The red nucleus located in the tegmentum of the mesencephalon is connected with the cerebral cortex, the cerebellum, brain stem nuclei, and the spinal cord. In the primate red nucleus, two subdivisions can be distinguished, i.e., the parvocellular and the magnocellular part. The crossed rubrobulbar and rubrospinal tract terminating on interneurons in the brain stem and spinal cord originates in the magnocellular part. The parvocellular part gives rise to the ipsilateral inferior olive (ten Donkelaar 1988; Massion 1967). In the adult human brain, the parvocellular part and its central tegmental tract are prominently developed, whereas the magnocellular part and its rubrospinal tract appear rudimentary in comparison to nonprimate brains.

In frontal Nissl-stained sections of the 6th or 7th gestational month, the red nucleus reveals an ovoid shape (Fig. 41). The magnocellular part appears like a semilunar shell around the inferior third of the parvocellular part. Both parts of the red nucleus are clearly separated from each other by a cell-sparse strip. The magnocellular part is

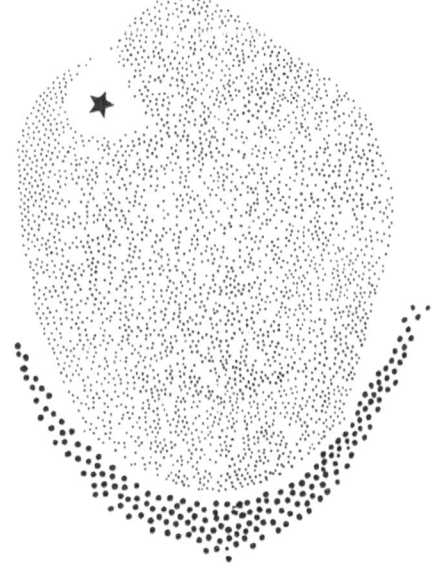

Fig. 41. Frontal Nissl-stained section passing through the red nucleus. *Small dots* represent the parvocellular part and *large dots* the magnocellular part. The *star* marks the area where the fasciculus retroflexus transverses the superomedial region of the red nucleus

characterized by darkly stained and densely packed neurons, showing a medium to large soma size. In the parvocellular part, a high packing density of small weakly stained neurons is observed.

11.2
Distribution Pattern of CB, CR, and PV in the Developing and Adult Red Nucleus

In CB immunopreparations of the 6th gestational month, the border between the two parts of red nucleus stands out clearly because the magnocellular part displays densely packed and intensely immunolabeled nerve cells, whereas the parvocellular part contains only a few ir fibers. At high magnification the ir neurons of the magnocellular part appear triangular or bipolar. The fibers of the parvocellular part do not show any preferential orientation (Ulfig and Chan 2001).

In CB immunosections of the perinatal brain, the magnocellular part also stands out conspicuously (Fig. 42). It harbors a high packing density of strongly immunolabeled neurons. The majority of these neurons show a large multipolar soma. Slender as well as stout dendrites spreading out in all directions arise from the polygonal cell bodies. A dense network of ir fibers is seen between the nerve cells. Intermingled between the multipolar neurons, a moderate number of medium-sized and small neurons, mostly revealing a bipolar or triangular shape, is observed (Fig. 43). The packing density of nerve cells is particularly high in the medial and lateral portions of the magnocellular part. In comparison to the findings in the brains of the 6th gestational month, the number of CB-ir neurons appears distinctly higher in the perinatal brain. On the whole, CB is the most abundant CaBP in the magnocellular part in the fetal and perinatal brain.

Within the parvocellular part, only fibrous immunoreactivity is present. These fibers run in bundles or singly and often cross each other. Fibers in the periphery of the parvocellular part are often in continuity with the magnocellular part.

In the adult brain, the number of CB-ir neurons in the magnocellular part is significantly reduced. In general, ir nerve cells are widely scattered. Mainly large multipolar neurons are found in groups of two or three or lying singly. The neuropil exhibits a few fibers and a moderate number of puncta. Within the parvocellular part, only weak punctate labeling is detected.

In CR immunopreparations of the 6th gestational month, ir neurons are found in high packing density within the magnocellular portion. These medium-sized to large neurons exhibit a bipolar, triangular, or multipolar shape. Moreover, a network of irregularly oriented CR-ir fibers is seen in the magnocellular portion.

In the parvocellular part, small CR-ir neurons with various shapes, which are only weakly immunolabeled, are mainly found in the medial third. In the lateral third, a moderate number of fibers oriented in various directions is observed.

Similar findings as in the 6th gestational month are seen in the perinatal brain (Fig. 44). The magnocellular part appears prominent because of densely packed and intensely immunolabeled large neurons, which mainly display a multipolar shape. In addition, a few small bipolar nerve cells are seen intermingled between the large neurons. The medial third of the parvocellular shows a moderate to high number of small CR-ir nerve cells displaying various soma shapes.

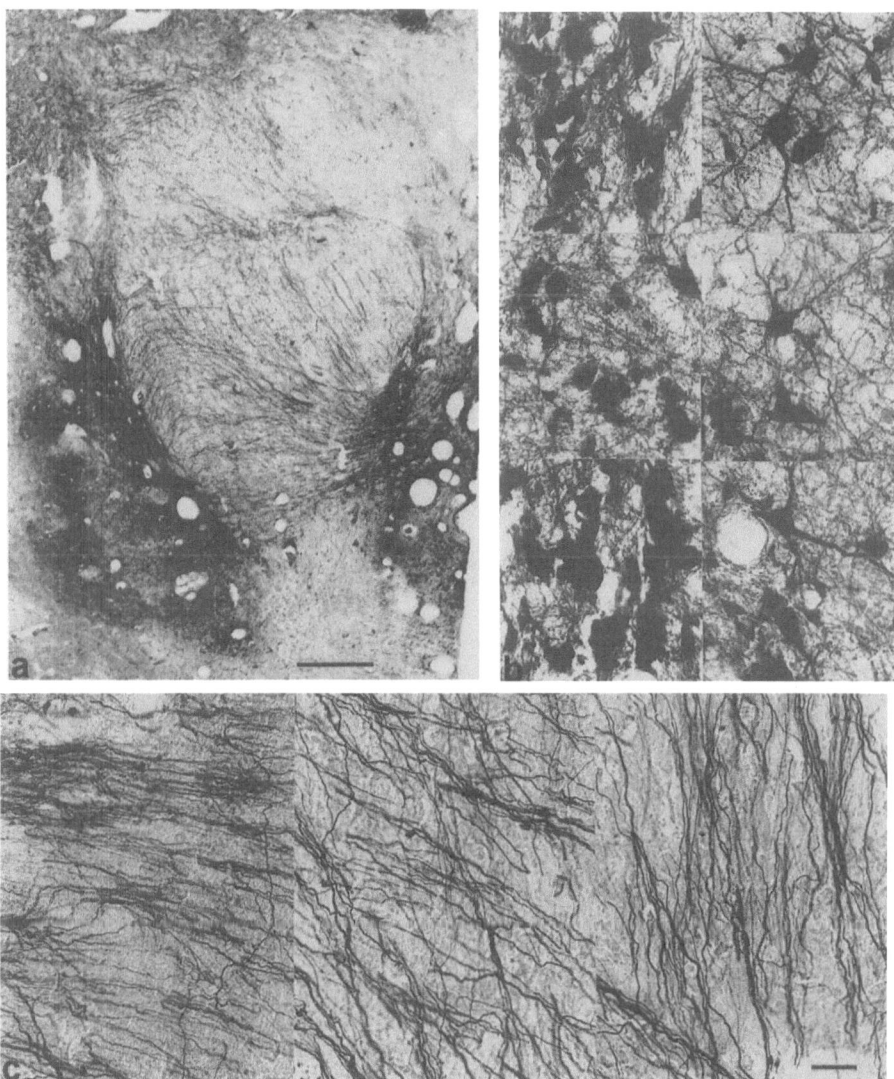

Fig. 42. a Frontal CB immunosection of the red nucleus, 2 weeks postnatally. Note the dense accumulation of ir structures in the magnocellular part; *l*, lateral; *m*, medial. **b** Mainly multipolar neurons being densely packed from the magnocellular part (enlargement taken from **a**, *stars*). **c** Fibers arranged in bundles or lying singly from the parvocellular part (enlargement from **a**, *asterisk*). Scale bars: **a**, 1 mm; **b** and **c**, 50 μm. (From Ulfig and Chan 2001, with permission)

In the adult brain, the magnocellular part contains only very few CR-ir nerve cells, mainly lying singly. Comparable to the results seen in the fetal and neonatal brain, the parvocellular part displays small neurons in its medial third. The middle and lateral thirds are nearly devoid of PV-ir structures.

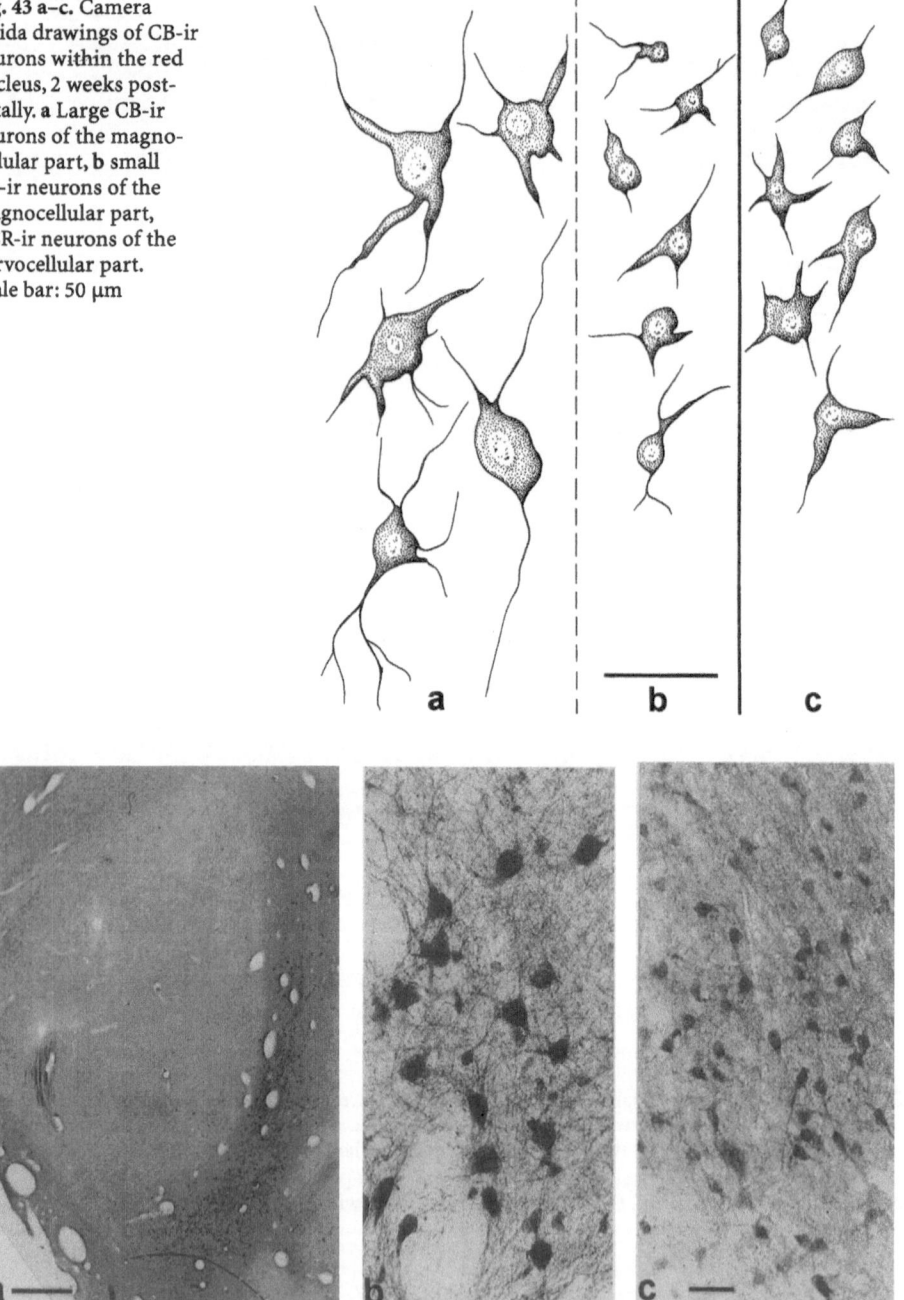

Fig. 43 a–c. Camera lucida drawings of CB-ir neurons within the red nucleus, 2 weeks postnatally. **a** Large CB-ir neurons of the magnocellular part, **b** small CB-ir neurons of the magnocellular part, **c** CR-ir neurons of the parvocellular part. Scale bar: 50 μm

Fig. 44. a Frontal CR immunosection of the red nucleus, 1st postnatal month; *l*, lateral; *m*, medial. **b** Medium-sized to large neurons from the magnocellular part (enlargement taken from **a**, *star*). **c** Neurons from the parvocellular part (enlargement taken from **a**, *asterisk*). Scale bars: **a**, 1 mm; **b** and **c**, 50 μm. (From Ulfig and Chan 2001, with permission)

In PV-immunopreparations of the 6th gestational month, mainly ir fibers are present in both parts of the red nucleus. Fibers are more densely packed in the magnocellular than in the parvocellular part. They can often be traced up to several millimeters. In the magnocellular part, a few PV-ir nerve cell bodies are seen.

In PV-immunosections of the perinatal brain, the parvocellular part harbors short ir fiber bundles and small ir cell bodies. The latter are mainly found in the medial third. At high magnification, the PV-ir fibers often reveal swellings. Within the magnocellular part only a few fibers are seen.

In the adult brain the parvocellular part contains fibrous and punctate PV immunostaining as well as a moderate number of small nerve cell bodies. These ir structures are evenly distributed within the parvocellular part.

The magnocellular part is devoid of PV-ir structures.

11.3
Relationship Between the Magnocellular and Parvocellular Parts

The features of the red nucleus as revealed by use of antisera against CaBPs indicate that the two parts, i.e., the magno- and parvocellular parts, are separate structures. In the fetal, perinatal, and adult brain, no merging of the two subdivisions is observed. In a study on the adult rat red nucleus using anti-CB and anti-PV, Hontanilla et al. (1995) found that a distinction between the parvocellular and magnocellular part is not visible in their immunosections. Comparison of the results obtained from the rat brain and those from the human brain yields distinct interspecies differences. Hontanilla et al. (1995) described PV- and CR-ir neurons of similar size in both parts of the red nucleus. Moreover, they observed a neuronal subpopulation which expresses both PV and CB. In the human developing as well as mature red nucleus, CB is only expressed in medium-sized to large neurons, whereas PV immunoreactivity is solely found in small neurons of the parvocellular part. Thus, in contrast to the rat, the human red nucleus consists of two parts that are clearly distinct with regard to cell morphology, neurochemical features, and architectonic delineations.

Interspecies differences also become obvious when comparing the specific connections of the two parts. In the rat red nucleus, neurons which project to the inferior olive send collaterals to the spinal cord (Kennedy 1990). A gradual reduction in collateralization is observed when ascending the phylogenetic scale (i.e., from reptiles to birds to mammals, and from opossum to rat to cat to monkey). This decrease in the number of collaterals is associated with an increase in facilities for independent movement of the upper limb (Huisman et al. 1982).

11.4
Neurochemical Characteristics of Rubric Nerve Cells

In many brain regions, PV has been shown to be colocalized with GABA (Celio 1990). Thus, the PV-ir nerve cells found in the developing and adult red nucleus may represent GABAergic neurons. This notion is in line with the finding of inhibitory GABAergic interneurons in the red nucleus. In the perinatal and adult red nucleus, a moderate number of PV-ir putative interneurons is present.

In the monkey brain, inhibitory synaptic input to rubrospinal neurons has been demonstrated (Ralston and Milroy 1992). PV-ir neurons can only be detected in the developing magnocellular part, and a gradual disappearance of this neuronal population takes place. These observations suggest that a transient neuronal population, i.e., inhibitory interneurons, exists within the magnocellular part.

Wang et al. (1996) provided evidence that not all rubrospinal neurons contain CB. This observation is supported by CB/CR-double-labelings of the perinatal red nucleus, which demonstrate that many magnocellular neurons express both CR and CB while others express CR only. Wang et al. (1996), moreover, showed that CB is restricted to the soma-dendritic compartment, whereas PV is mainly found in axons. In accordance with these findings, a high number of PV-ir fibers is seen in the developing human red nucleus, whereas the number of PV-ir neurons is relatively small.

PV immunopreparations reveal PV accumulations within axonal swellings, suggesting that PV may be involved in the transportation of synaptic vesicles.

11.5
The Magnocellular Part of the Red Nucleus
in the Adult Human Brain

Early cytoarchitectonic studies have shown that the magnocellular part is rudimentary in the adult human brain. Magnocellular neurons appear as scattered cells or occur in groups of two to five (Hatschek 1907; von Monakow 1909; Winkler 1929). The results obtained from CB- and CR immunopreparations of the human adult red nucleus clearly support these data.

In adult primate brains (including humans), the number of fibers making up the rubrospinal tract is small and these fibers cannot be traced caudal to the upper cervical segments (Voogd et al. 1998). Comparative studies have provided evidence that, when moving upwards on the evolutionary scale, the relative size of the magnocellular part and the rubrospinal tract becomes smaller (Nathan and Smith 1982).

11.6
The Magnocellular Part of the Red Nucleus
in the Developing Human Brain

Comparing the fetal or perinatal with the adult red nucleus, it is obvious that the magnocellular part is conspicuously prominent relative to the parvocellular part in the human developing brain. Most probably, this prominence can be correlated to a well-developed rubrospinal projection during development. The exact mechanism underlying the reduction of the magnocellular part during postnatal development is so far not known. When comparing the magnocellular part at various stages of development and in the mature brain, one might assume that the ontogeny follows the phylogeny.

It is tempting to assume that the prominence of the magnocellular part represents a morphological substrate for a specific transitory pattern of motor behavior. Thus, an adaption to the environment may be reflected by the relative size of the two subdivisions of the red nucleus. The strong flexor tone observed in the extremities in late fetal and postnatal life may be correlated with a well-developed rubrospinal projection. A dominance of the parvocellular part over the magnocellular part, as observed with proceeding maturation may be correlated with the development of independent arm-hand movements (Massion 1967), just as one might expect when climbing up the evolutionary scale.

11.7
Significance of the Prominent Magnocellular Part for Neuropediatrics

Periventricular leukomalacia mainly occurs in premature infants and represents a bilateral white matter necrosis near the external angles of the lateral ventricles. The acute hypoxic-ischemic insult leading to the necrosis is acquired around birth but neurological disability, i.e., spastic paresis of the lower limbs, is only observed later in life (Friede 1989; Volpe 1995). The mechanism underlying the delay in occurrence of the major sequela is so far not understood. By bringing together neuroanatomical, neuroembryonic as well experimental data on the red nucleus and clinical observations on periventricular leukomalacia, a conceivable explanation can be deduced. After lesioning the corticospinal tract in adult rats, recovery of function occurs to a considerable extent. Most probably, the underlying mechanism is a switch from the corticospinal to the rubrospinal tract (Kennedy 1990). After lesions to the corticospinal tract, error signals are sent to the cerebral as well as cerebellar cortex, which both in turn project to the parvocellular part. Corrective signals subsequently are sent through the rubro-olivary tract (arising in the parvocellular part) and then through the olivocerebellar tract. Finally, these corrective signals travel through the cerebellorubral pathway, which activates rubrospinal neurons. Via these circuitries a switching from the corticospinal to the rubrospinal tract is achieved.

As the rubrospinal projection is negligible in adult humans, recovery from stroke cannot occur via the aforementioned switching. In contrast, in the newborn infant, a rubrospinal projection exists. Thus lesions of the corticospinal fibers for the leg (in periventricular leukomalacia) are not detectable in clinical examination. In accordance with these hypotheses, infants display an increase in flexor tone 6–10 weeks after the lesion. This increase may result from a higher activity in the rubrospinal tract which is known to control flexor muscle tone. The late and gradual occurrence of spastic diplegia may be linked to a continuous reduction of rubrospinal fibers during postnatal development.

12 Summary

Recent studies have demonstrated that antibodies against the calcium-binding proteins (CaBPs) parvalbumin (PV), calbindin (CB), and calretinin (CR) are appropriate tools for demonstrating transient features and developmental changes of human fetal brain organization as well as for detecting specific alterations in pathologically altered specimens. CB and CR are abundantly expressed in various nerve cell types of the subplate in the second half of gestation. The subplate being an outstandingly wide zone subjacent to the cortical plate, it is a "waiting compartment" for various cortical afferents that reside here prior to entering the cortical plate. The cortical plate (future layers II–VI of the cerebral cortex) contains only CR-ir neurons until the 6th gestational month. In the 7th and 8th month, cortical CB- and PV-ir interneurons are observed in deeper portions of the cortical plate. Cajal-Retzius cells of layer I are CR-immunolabeled from the 4th month onwards. Fetal hydrocephalus causes severe alterations of CB- and PV-ir neurons in the subplate and the cortical plate: shrinkage of ir neurons, loss of process labeling and in most severe cases, entire loss of immunolabeling. Such alterations, which cannot be detected in Nissl-stained sections, indicate distinct impairment of neuronal function.

The ganglionic eminence being a prominent part of the telencephalic proliferative zone persists nearly throughout the entire fetal period. Between 16 and 24 weeks of gestation, CR-ir cells are found in the center and, in a higher number, in the periphery, i.e., the mantle zone, of the ganglionic eminence. The mantle zone also exhibits CB-ir cells. These observations support experimental data showing that CR-ir precursor cells leave the ganglionic eminence to migrate towards the cerebral cortex. The CR- and CB-ir neurons of the mantle zone most probably represent an intermediate target for outgrowing axons. This notion is supported by the observation that SNAP (synaptosomal associated protein) 25-ir fibers coming from the intermediate zone terminate upon CR-ir cells in the mantle zone.

Within the amygdaloid complex, immature, migrating CR- or CB-ir neurons are observed in the 5th and 6th gestational month. In the 8th and 9th month, anti-CR and anti-CB mark different subsets of interneurons as well as a small proportion of pyramidal projection neurons. The different subsets of interneurons are likely to be functionally different with regard to their connectivities. Considering studies in the literature, it is obvious that CR is transiently expressed in pyramidal cells. Moreover, diffuse (neuropil) CB and CR immunolabeling, which is found in different intensities in the various amygdaloid subdivisions, displays distinct redistribution during development, an observation indicating reorganization of afferent inputs.

The sequential arrival of various afferent fiber systems in the two compartments of the striatum (patch and matrix compartment) is reflected by changing patterns of diffuse CB immunolabeling: During the second half of gestation, the patches are labeled and postnatally a changeover to matrix labeling is seen.

The thalamic reticular complex reveals prominent transient features seen in PV and CR immunopreparations. Four subdivisions become obvious: the main portion, the perireticular nucleus, the medial subnucleus, and the pregeniculate nucleus. The PV- and CR-ir perireticular nucleus, not visible in the mature brain, is a distinct fetal entity located within the internal capsule. The main portion of the reticular complex is much more prominent in the fetus than in the adult and displays transitory CR expression. The most probable developmental role of the reticular complex is to provide guiding cues for outgrowing axons from or into the dorsal thalamus.

The basal nucleus of Meynert and the hypothalamic tuberomamillary nucleus both provide extrathalamic projections to the cerebral cortex. The sequential differentiation of the two nuclei can be demonstrated using anti-CB and anti-PV. The basal nucleus strongly expresses CB and appears to be mature distinctly earlier than the PV-ir tuberomamillary nucleus.

Antisera against CaBPs clearly demonstrate that the magnocellular part of the red nucleus located in the mesencephalic tegmentum is outstanding in the fetal and perinatal brain and inconspicuous in the adult. In particular, CB is the most abundant CaBP in this portion of the red nucleus. The dominance of the magnocellular part over the parvocellular part may be a substrate for a specific transitory pattern of motor behavior.

On the whole, CaBPs mark the transient architectonic organization of the brain, which is involved in the establishment of transitory neuronal circuitries. The latter are essential for the formation of mature projections. Detailed data on the normal organization of the transient structures are required for the evaluation of alterations occurring in the fetal and perinatal brain. The transient structures are sites of predilection for alteration caused by hypoxia-ischemia, hemorrhage, or hydrocephalus.

References

Aggleton JP (1993) The contribution of the amygdala to the normal and abnormal emotional states. Trends Neurosci 16:328–333

Allendoerfer KL, Shatz CJ (1994) The subplate, a transient neocortical structure: its role in the development of connections between thalamus and cortex. Annu Rev Neurosci 17:185–218

Al-Mohanna FA, Cave J, Bolsover SR (1992) A narrow window of intracellular calcium concentration is optimal for neurite outgrowth in rat sensory neurones. Brain Res Dev Brain Res 70:287–290

Amaral DG, Price JL, Pitkänen A, Carmichael ST (1992) Anatomical organization of the primate amygdaloid complex. In: Aggleton JP (ed) The amygdala. Wiley-Liss Inc., New York, pp 1–66

Andressen C, Blumcke I, Celio MR (1993) Calcium binding proteins: selective markers of nerve cells. Cell Tissue Res 271:181–208

Auer RN, Benveniste H (1997) Hypoxia and related conditions. In: Graham DI, Lantos PL (eds). Greenfield's neuropathology. Arnold, London, pp 263–314

Baimbridge KG, Celio MR, Rogers JH (1992) Calcium-binding proteins in the nervous system. Trends Neurosci 15:303–308

Bayer SA, Altmann S (1991) Neocortical development. Raven Press, New York

Berchtold MW (1989) Parvalbumin genes from human and rat are identical in intron/exon organization and contain highly homologous regulatory elements and coding sequences. J Mol Biol 210:417–427

Bergmann M, Lahr G, Mayerhofer A, Gratzl M (1991) Expression of synaptophysin during the prenatal development of the rat spinal cord: correlation with basic differentiation processes of neurons. Neuroscience 42:569–582

Borhegyi Z, Leranth C (1997) Distinct substance P- and calretinin-containing projections from the supramamillary area to the hippocampus in rats; a species difference between rats and monkeys. Exp Brain Res 115:369–374

Braak H, Braak E (1983) Neuronal types in basolateral amygdaloid nuclei of man. Brain Res Bull 11:349–365

Brown MC, Hopkins WG, Keynes RS (1991) Essentials of neural development. Cambridge University Press, Cambridge

Brown MC, Keynes RS, Lumsden A (2001) The developing brain. Oxford University Press, Oxford New York

Celio MR (1990) Calbindin D28k and parvalbumin in the rat nervous system. Neuroscience 35:375–475

Celio MR, Heizmann CW (1981) Calcium-binding protein parvalbumin as a neuronal marker. Nature 293:300–302

Celio MR, Norman AW (1985) Nucleus basalis Meynert neurons contain the vitamin D-induced calcium-binding protein (Calbindin-D28k). Anat Embryol 173:143–148

Chan WY, Kostovic I, Takashima S, Feldhaus C, Stoltenburg-Didinger G, Verney C, Yew D, Ulfig N (2002) Normal and abnormal development of the human cerebral cortex. Neuroembryology 1: 78–90

Charvet I, Hemming FJ, Feuerstein C, Saxod R (1998) Mosaic distribution of chondroitin and keratan sulphate in the developing rat striatum: possible involvement of proteoglycans in the organization of the nigrostriatal system. Brain Res Dev Brain Res 109:229–244

Cheung WY (1980) Calmodulin plays a pivotal role in cellular regulation. Science 207:19–27

Christakos S, Friedlander EJ, Frandsen BR, Norman AW (1979) Studies on the mode of action of calciferol. XIII. Development of radioimmunoassay for vitamin D-dependent chick intestinal calcium-binding protein and tissue distribution. Endocrinology 104:1495–1503

Chun JJ, Shatz CJ (1989) Interstitial cells of the adult neocortical white matter are the remnant of the early generated subplate neuron population. J Comp Neurol 282:555–569

Chun JJ, Nakamura MJ, Shatz CJ (1987) Transient cells of the developing mammalian telencephalon are peptide-immunoreactive neurons. Nature 325:617–620

Clemence AE, Mitrofanis J (1992) Cytoarchitectonic heterogeneities in the thalamic reticular nucleus of cats and ferrets. J Comp Neurol 322:167–180

Cline HAT, Tsien RW (1991) Glutamate-induced increases in intracellular Ca2+ in cultured frog tectal cells mediated by direct activation of NMDA receptor channels. Neuron 6:259–267

Coghlan VM, Hausken ZE, Scott JD (1995) Subcellular targeting of kinases and phosphatases by association with bifunctional anchoring proteins. Biochem Soc Trans 23:592–596

Cornwall J, Cooper JD, Phillipson OT (1990) Projections to the rostral reticular thalamic nucleus in the rat. Exp Brain Res 80:157–171

D'Arcangelo G, Miao GG, Chen SC, Soares HD, Morgan JI, Curran T (1995) A protein related to extracellular matrix proteins deleted in the mouse mutant reeler. Nature 374:719–723

Dash PK, Karl KA, Colicos MA, Prywes R, Kandel ER (1991) cAMP response element-binding protein is activated by Ca2+/calmodulin as well as cAMP-dependent protein kinase. Proc Natl Acad Sci U S A 88:5061–5065

De Leon M, Convenas R, Narvaez JA, Aguirre JA, Gonzales-Baron S (1994) Distribution of Calbindin D-28k-immunoreactivity in the cat brainstem. Arch Ital Biol 132:229–241

De Viragh PA, Haglid KG, Celio MR (1989) Parvalbumin increases in the caudate putamen of rats with vitamin D hypervitaminosis. Proc Natl Acad Sci U S A 86:3887–3890

Del Bigio MR (1993) Neuropathological changes caused by hydrocephalus. Acta Neuropathol 85:573–85

Earle KL, Mitrofanis J (1996) Genesis and fate of the perireticular thalamic nucleus during early development. J Comp Neurol 367:246–263

Ellis JH, Richards DE, Rogers JH (1991) Calretinin and calbindin in the retina of the developing chick. Cell Tissue Res 264:197–208

Enderlin S, Norman AW, Celio MR (1987) Ontogeny of the calcium binding protein calbindin D-28k in the rat nervous system. Anat Embryol 177:15–28

Faissner A, Schachner M (1995) Tenascin and janusin: glial recognition molecules involved in neural development. In: Kettenmann H, Ransom BR (eds.) Neuroglia. Oxford University Press, New York, pp 411–426

Fallon JH, Ciofi P (1992) Distribution of monoamines within the amygdala. In: Aggleton JP (ed) The amygdala. Wiley-Liss Inc., New York, pp 97–114

Fonseca M, del Rio JA, Martinez A, Gomez S Soriano E (1995) Development of calretinin immunoreactivity in the neocortex of the rat. J Comp Neurol 361:177–192

Frassoni C, Arcelli P, Selvaggio M, Spreafico R (1998) Calretinin immunoreactivity in the developing thalamus of the rat: a marker of early generated thalamic cells. Neuroscience 83:1303–1214

Friauf E, McConnell SK, Shatz CJ (1990) Functional synaptic circuits in the subplate during fetal and early postnatal development of cat visual cortex. J Neurosci 10:2601–2613

Friede RL (1989) Developmental neuropathology. Springer, New York

Gerfen CR, Baimbridge KG, Miller JJ (1985) The neostriatal mosaic: compartmental distribution of calcium-binding protein and parvalbumin in the basal ganglia of the rat and monkey. Proc Natl Acad Sci U S A 82:8780–8784

Ghosh A, Shatz CJ (1992) Involvement of subplate neurons in the formation of ocular dominance columns. Science 255:1441–1443

Ghosh A, Carnahan J, Greenberg ME (1994) Requirement for BDNF in activity-dependent survival of cortical neurons. Science 263:1618–1623

Glantz SB, Amat JA, Rubin CS (1992) cAMP signaling in neurons: pattern of neuronal expression and intracellular localization for a novel protein, AKAP 150, that anchors the regulatory subunit of cAMP-dependent protein kinase II beta. Mol Biol Cell 3:1215–1228

Gomez TM, Snow DM, Letourneau PC (1995) Characterization of spontaneous calcium transient in nerve growth cones and their effect on growth cone migration. Neuron 14:1233–1246

Graybiel AM (1984) Correspondence between the dopamine islands and striosomes of the mammalian striatum. Neuroscience 13:1157–1187

Graybiel AM (1990) Neurotransmitters and neuromodulators in the basal ganglia. Trends Neurosci 13:244–254

Graybiel AM, Ragsdale CW Jr (1980) Clumping of acetylcholinesterase activity in the developing striatum of the human fetus and young infant. Proc Natl Acad Sci U S A 77:1214–1218

Graybiel AM, Ragsdale CW Jr, Yoneoka ES, Elde RP (1981) An immunohistochemical study of enkephalins and other neuropeptides in the striatum of the cat with evidence that the opiate peptides are arranged to form mosaic patterns in register with the striosomal compartments visible by acetylcholinesterase staining. Neuroscience 6:377–397

Gupta RS, Dudani AK (1989) Mechanism of action of antimitotic drugs: a new hypothesis based on the role of cellular calcium. Med Hypotheses 28:57–69

Hammerschlag R, Dravid AR, Chiu AY (1975) Mechanism of axonal transport: a proposed role for calcium ions. Science 188:273–275

Hatschek R (1907) Zur vergleichenden Anatomie des Nucleus ruber tegmenti. Arbeiten aus dem neurologischen Institut an der Wiener Universität 15:89–136

Heizmann CW (1984) Parvalbumin, an intracellular calcium-binding protein; distribution, properties and possible roles in mammalian cells. Experientia 40:910–921

Heizmann CW, Braun K (1992) Changes in Ca(2+)-binding proteins in human neurodegenerative disorders. Trends Neurosci 15:259–264

Hendrickson AE, Van Brederode JF, Mulligan KA, Celio MR (1991) Development of the calcium-binding protein parvalbumin and calbindin in monkey striate cortex. J Comp Neurol 307:626–646

Hinrichsen RD, Burgess-Cassler A, Soltvedt BC, Hennessey T, Kung C (1986) Restoration by calmodulin of a Ca2+-dependent K+ current missing in a mutant of paramecium. Science 232:503–506

Hontanilla B, Parent A, Gimenez-Amaya JM (1995) Heterogeneous distribution of neurons containing calbindin D-28k and/or parvalbumin in the rat red nucleus. Brain Res 696:121–126

Huisman AM, Kuypers HG, Verburgh CA (1982) Differences in collateralization of the descending spinal pathways from red nucleus and other brain stem cell groups in cat and monkey. Prog Brain Res 57:185–217

Humphrey T (1968) The development of the human amygdala during early embryonic life. J Comp Neurol 132:135–165

Huntley GW, Jones EG (1990) Cajal-Retzius neurons in developing monkey neocortex show immunoreactivity for calcium binding proteins. J Neurocytol 19:200–212

Jahn R, Schiebler W, Ouimet C, Greengard P (1985) A 38,000-dalton membrane protein (p38) present in synaptic vesicles. Proc Natl Acad Sci U S A 82:4137–4141

Jiang M, Swann JW (1997) Expression of calretinin in diverse neuronal populations during development of rat hippocampus. Neuroscience 81:1137–1154

Jones EG (1985) The thalamus. Plenum Press, New York

Kahle W (1969) Die Entwicklung der menschlichen Großhirnhemisphäre. Springer-Verlag, Heidelberg

Kater SB, Mattson MP, Cohan C, Connor J (1988) Calcium regulation of the neuronal growth cone. Trends Neurosci 11:315–321

Kennedy PR (1990) Corticospinal, rubrospinal and rubro-olivary projections: unifying hypothesis. Trends Neurosci 13:474–479

Kiss J, Csaki A, Bokor H, Kocsis K, Szeiffert G (1997) Topographic localization of calretinin, calbindin, VIP, substance P, CCK and metabotropic glutamate receptor immunoreactive neurons in the supramamillary and related areas of the rat. Neurobiology 5:361–388

Klauck TM, Scott JD (1995) The postsynaptic density: a subcellular anchor for signal transduction enzymes. Cell Signal 7:747–757

Klauck TM, Faux MC, Labudda K, Langeberg LK, Jaken S, Scott JD (1996) Coordination of three signaling enzymes by AKAP79, a mammalian scaffold protein. Science 271:1589–1592

Kodoma S (1929) Über die sogenannten Basalganglien 2 (Morphogenetische und pathologisch-anatomische Untersuchungen). Schweizer Arch Neurol Psychiat 18:179–246

Köhler C, Swanson LW, Haglund L, Wu JY (1985) The cytoarchitecture, histochemistry and projections of the tuberomamillary nucleus in the rat. Neuroscience 16:85–110

Komuro H, Rakic P (1996) Intracellular Ca2+ fluctuations modulate the rate of neuronal migration. Neuron 17:275–285

Kordower JH, Mufson EJ (1992) Nerve growth factor receptor-immunoreactive neurons within the developing human cortex. J Comp Neurol 323:25–41

Kordower JH, Rakic P (1990) Neurogenesis of the magnocellular basal forebrain nuclei in the rhesus monkey. J Comp Neurol 291:637–653

Kostovic I (1986) Prenatal development of nucleus basalis complex and related fiber systems in man: a histochemical study. Neuroscience 17:1047–1077

Kostovic I (1990a) Structural and histochemical reorganization of the human prefrontal cortex during perinatal and postnatal life. Prog Brain Res 85:223–239

Kostovic I (1990b) Zentralnervensystem In: Hinrichsen (ed) Humanembryologie. Springer-Verlag, Heidelberg, pp 381–448

Kostovic I, Rakic P (1990) Development history of the transient subplate zone in the visual and somatosensory cortex of the macaque monkey and human brain. J Comp Neurol 297:441–470

Kostovic I, Lukinovic N, Judas M, Bogdanovic N, Mrzljak L, Zecevic N, Kubat M (1989) Structural basis of the developmental plasticity in the human cerebral cortex: the role of the transient subplate zone. Metab Brain Dis 4:17–23

Kostovic I, Stefulj-Fucic A, Mrzljak L, Jukic S, Delalle I (1991) Prenatal and perinatal development of the somatostatin-immunoreactive neurons in the human prefrontal cortex. Neurosci Lett 124:153–156

Kreutzberg GW, Blakemore WF, Graeber MB (1997) Cellular pathology of the central nervous system. In: Graham DI, Lantos PL (eds) Greenfield's neuropathology. Arnold, London, pp 85–156

Krushel LA, Connolly JA, van der Kooy D (1989) Pattern formation in the mammalian forebrain: patch neurons from the rat striatum selectively reassociate in vitro. Brain Res Dev Brain Res 47:137–142

Lammers GJ, Gribnau AA, ten Donkelaar HJ (1980) Neurogenesis in the basal forebrain in the Chinese hamster (cricetulus griseus). II. Site of neuron origin: morphogenesis of the ventricular ridges. Anat Embryol 158:193–211

Leclerc N, Beesley PW, Brown I, Colonnier M, Gurd JW, Paladino T, Hawkes R (1989) Synaptophysin expression during synaptogenesis in the rat cerebellar cortex. J Comp Neurol 280:197–212

Letinic K, Kostovic I (1996a) Transient neuronal population of the internal capsule in the developing human cerebrum. Neuroreport 7:2159–2162

Letinic K, Kostovic I (1996b) Transient patterns of calbindin-D28k expression in the developing striatum of man. Neurosci Lett 220:211–214

Letinic K, Kostovic I (1997) Transient fetal structure, the gangliothalamic body connects telencephalic germinal zone with all thalamic regions in the developing human brain. J Comp Neurol 384:373–395

Linden DC, Guillery RW, Cucchiaro J (1981) The dorsal lateral geniculate nucleus of the normal ferret and its postnatal development. J Comp Neurol 203:189–211

Liu FC, Graybiel AM (1992a) Heterogeneous development of calbindin-D28K expression in the striatal matrix. J Comp Neurol 320:304–322

Liu FC, Graybiel AM (1992b) Transient calbindin-D28K-positive systems in the telencephalon: ganglionic eminence, developing striatum and cerebral cortex. J Neurosci 12:674–690

Lizier C, Spreafico R, Battaglia G (1997) Calretinin in the thalamic reticular nucleus of the rat: distribution and relationship with ipsilateral and contralateral efferents. J Comp Neurol 13:217–233

Lumsden A, Gulisano M (1997) Neocortical neurons: where do they come from? Science 278:474–476

Macchi G (1951) The ontogenetic development of the olfactory telencephalon in man. J Comp Neurol 95:245–305

Marchand R, Lajoie L (1986) Histogenesis of the striopallidal system in the rat. Neurogenesis of its neurons. Neuroscience 17:573–590

Marin-Padilla M (1992) Ontogenesis of the pyramidal cell of the mammalian neocortex and developmental cytoarchitectonics: a unifying theory. J Comp Neurol 321:223–240

Marin-Padilla M (1998a) Cajal-Retzius cells and the development of the neocortex. Trends Neurosci 21:64–71

Marin-Padilla M (1998b) Ontogenesis of the pyramidal cell of the mammalian neocortex and developmental cytoarchitectonics: a unifying theory. J Comp Neurol 321:223–240

Masliah E, Terry RD, Alford M, DeTeresa R (1990) Quantitative immunohistochemistry of synapto-physin in human neocortex: an alternative method to estimate density of presynaptic terminals in paraffin sections. J Histochem Cytochem 38:837–844

Massion J (1967) The mammalian red nucleus. Physiol Rev 47:383–436

Mattson MP, Kater SB (1987) Calcium regulation of neurite elongation and growth cone motility. J Neurosci 7:4034–4043

McAllister JP 2nd, Maugans TA, Shah MV, Truex RC Jr (1985) Neuronal effects of experimentally induced hydrocephalus in newborn rats. J Neurosurg 63:776–783

McConnell SK, Ghosh A, Shatz CJ (1989) Subplate neurons pioneer the first axon pathway from the cerebral cortex. Science 245:978–982

McDonald AJ (1992) Cell types and intrinsic connections of the amygdala. In: Aggleton JP (ed) The amygdala. Wiley-Liss Inc., New York, pp 67–96

Mehra RD, Hendrickson AE (1993) A comparison of the development of neuropeptide and MAP2 immunocytochemical labelling in the macaque visual cortex during pre- and postnatal develop-ment. J Neurobiol 24:101–124

Meinecke DL, Rakic P (1992) Expression of GABA and GABAA receptors by neurons of the subplate zone in developing primate occipital cortex: evidence for transient local circuits. J Comp Neurol 317:91–101

Metin C, Godement P (1996) The ganglionic eminence may be an intermediate target for corticofugal and thalamocortical axons. J Neurosci 16:3219–3235

Metin C, Deleglise D, Serafini T, Kennedy TE, Tessier-Lavigne M (1997) A role for netrin-1 in the guidance of cortical efferents. Development 124:5063–5074

Meyer G, Gonzales-Hernandez T (1993) Developmental changes in layer I of the human neocortex during prenatal life: a DiI-tracing and AChE and NADPH-d histochemistry study. J Comp Neurol 338:317–336

Miettinen R, Gulyas AI, Baimbridge KG, Jacobowitz DM, Freund TF (1992) Calretinin is present in non-pyramidal cells of the rat hippocampus. II. Co-existence with other calcium binding proteins and GABA. Neuroscience 48:29–43

Miller RJ (1991) The control of neuronal Ca2+ homeostasis. Prog Neurobiol 37:255–285

Mitrofanis J (1994) Development of the thalamic reticular nucleus in ferrets with special reference to the perigeniculate and perireticular cell groups. Eur J Neurosci 6:253–263

Mitrofanis J, Guillery RW (1993) New views of the thalamic reticular nucleus in the adult and the developing brain. Trends Neurosci 16:240–245

Miyajima M, Sato K, Arai H (1996) Choline acetyltransferase, nerve growth factor and cytokine levels are changed in congenitally hydrocephalic HTX rats. Pediatr Neurosurg 24:1–4

Molnar Z (1998) Development of thalamocortical connections. Springer, Berlin Heidelberg New York

Mrzljak L, Uylings HB, Kostovic I, Van Eden CG (1988) Prenatal development of neurons in the human prefrontal cortex: I. A quantitative Golgi study. J Comp Neurol 271:355–386

Nathan PW, Smith MC (1982) The rubrospinal and central tegmental tracts in man. Brain 105:223–269

Nikolic I, Kostovic I (1986) Development of the lateral amygdaloid nucleus in the human fetus: transient presence of discrete cytoarchitectonic units. Anat Embryol 174: 355–360

Oudega M, Lakke EA, Marani E, Thomeer RT (1993) Development of the rat spinal cord: immuno- and enzyme histochemical approaches. Adv Ana Embryol Cell Biol 129:1–166

Ovtscharoff W, Bergmann M, Marqueze-Pouey B, Knaus P, Betz H, Grabs D, Panula P, Airaksinen MS, Pirvola U, Kotilainen E (1990) A histamine-containing neuronal system in human brain. Neuroscience 34:127–132

Panula P, Airaksinen MS (1991) The histaminergic neuronal system as revealed with antisera against histamine. In: Watanabe T, Wanda H (eds) Histaminergic neurons, their structure and functions. CRC-Press, Boca Raton, pp 27–144

Panula P, Pirvola U, Airaksinen MS (1989) Histamine-immunoreactive nerve fibers in the rat brain. Neuroscience 28:585–610

Parè D, Hazrati LN, Parent A, Steriade M (1990) Substantia nigra pars reticulata projects to the reticular thalamic nucleus of the cat: a morphological and electrophysiological study. Brain Res 535:139–146

Parmentier M (1980) The human calbindins: cDNA and gene cloning. Adv Exp Med Biol 255:233–240

Parmentier M, Passage E, Vassard G, Mattei MG (1991) The human calbindin D28k (CALB1) and calretinin (CALB2) genes are located at 8q21.3--q22.1 and 16q22-q23, respectively, suggesting a common duplication with the carbonic anhydrase isozyme loci. Cytogenet Cell Genet 57:42–43

Parnavelas JG (2000) The origin and migration of cortical neurones: new vistas. Trends Neurosci 23:126–131

Pasteels B, Miki N, Hatakenaka S, Pochet R (1987) Immunohistochemical cross-reactivity and electrophoretic comigration between calbindin D-27kDa and visinin. Brain Res 412:107–113

Paul A, Ulfig N (1998) Lectin staining in the basal nucleus (Meynert) and the hypothalamic tubero-mamillary nucleus of the developing human prosencephalon. Anat Rec 252:149–158

Pinto Lord MC, Caviness VS Jr (1979) Determinants of cell shape and orientation: a comparative Golgi analysis of cell-axon interrelationships in the developing neocortex of normal and reeler mice. J Comp Neurol 187:49–69

Pitkänen A, Amaral DG (1993) Distribution of calbindin-D28k immunoreactive in the monkey temporal lobe: the amygdaloid complex. J Comp Neurol 331:199–224

Pochet R, Parmentier M, Lawson DE, Pasteels JL (1985) Rat brain synthesizes two 'vitamin D-dependent' calcium-bindings proteins. Brain Res 345:251–256

Raimondi AJ, Soare P (1974) Intellectual development in shunted hydrocephalic children. Am J Dis Child 127:664–671

Rakic P (1974) Neurons in rhesus monkey visual cortex: systematic relation between time of origin and eventual disposition. Science 183:425–427

Rakic P (1988) Specification of cerebral cortical areas. Science 241:170–176

Rakic P (1990) Principles of neural cell migration. Experientia 46:882–891

Rakic P (1995) Radial glial cells: Scaffolding for brain construction. In: Kettenmann H, Ranson BR (eds) Neuroglia. Oxford University Press, New York, pp 746–762

Rakic P, Sidman RL (1969) Telencephalic origin of pulvinar neurons in the fetal human brain. Z Anat Entwicklungsgesch 129:53–82

Ralston DD, Milroy AM (1992) Inhibitory synaptic input to identified rubrospinal neurons in Macaca fascicularis: an electron microscopic study using a combined immuno-GABA-gold technique and the retrograde transport of WGA-HRP. J Comp Neurol 320:97–109

Ramon y Cajal S (1911) Histologie des systèmes nerveux de l'homme et des vertébrés. Maloine, Paris

Rancharan EJ, Guillery RW (1997) Membrane specializations in the developmentally transient perireticular nucleus of the rat. J Comp Neurol 380:435–448

Reisert I, Gratzl M (1993) Ontogeny of synaptophysin and synaptoporin in the central nervous system: differential expression in striatal neurons and their afferents during development. Brain Res Dev Brain Res 72:219–225

Retzius G (1893) Die Cajal'schen Zellen der Grosshirnrinde beim Menschen und bei Säugetieren. Biol Untersuch 5:1–9

Rizvi TA, Ennis M, Behbehani MM, Shipley MT (1991) Connections between the central nucleus of the amygdala and the midbrain periaqueductal gray: topography and reciprocity. J Comp Neurol 303:121–131

Royce GJ, Bromley S, Gracco C (1991) Subcortical projections to the centromedian and parafascicular thalamic nuclei in the cat. J Comp Neurol 306:129–155

Rubin CS (1994) A kinase anchor proteins and the intracellular targeting of signals carried by cyclic AMP. Biochim Biophys Acta 1224:467–479

Sainte-Rose C (1993) Hydrocephalus in childhood. In: Apuzzo MLJ (ed) Brain surgery: complication avoidance and management, vol. 2,. Churchill Livingstone, New York, pp 890–926

Saper CB (1990) Hypothalamus. In: Paxinos G (ed) The human nervous system. Academic Press, New York, pp 389–413

Sarthy PV, Bacon W (1985) Developmental expression of a synaptic vesicle-specific protein in the rat retina. Dev Biol 112:284–291

Schierle GS, Gander JC, D'Orlando C, Ceilo MR, Vogt Weisenhorn DM (1997) Calretinin-immunoreactivity during postnatal development of the rat isocortex: a qualitative and quantitative study. Cereb Cortex 7:130–142

Setzer M, Ulfig N (1999) Differential expression of calbindin and calretinin in the human fetal amygdala. Microsc Res Tech 46:1–17

Solbach S, Celio MR (1991) Ontogeny of calcium binding protein parvalbumin in the rat nervous system. Anat Embryol 184:103–124

Song DD, Harlan RE (1994) Genesis and migration patterns of neurons forming the patch and matrix compartments of the rat striatum. Brain Res Dev Brain Res 83:233–245

Sorvari H, Soininen H, Pitkänen A (1996a) Calretinin-immunoreactive cells and fibers in the human amygdaloid complex. J Comp Neurol 369:188–208

Sorvari H, Soininen H, Pitkänen A (1996b) Calbindin-D28k-immunoreactive cells and fibres in the human amygdaloid complex. Neuroscience 75:421–443

Sorvari H, Miettinen R, Soininen H, Paljarvi L, Karkola K, Pitkänen A (1998) Calretinin immunoreactive terminals make synapses on calbindin D28k-immunoreactive neurons in the lateral nucleus of the human amygdala. Brain Res 783:355–358

Spitzer NC (1994) Spontaneous Ca2+spikes and waves in embryonic neurons: signaling systems for differentiation. Trends Neurosci 17:115–118

Steindler DA, O`Brien TF, Cooper NG (1988) Glycoconjugate boundaries during early postnatal development of the neostriatal mosaic. J Comp Neurol 267:357–369

Stichel CC, Singer W, Heizmann CW, Norman AW (1987) Immunohistochemical localization of calcium-binding proteins, parvalbumin and calbindin-D28k, in the adult and developing visual cortex of cats: a light and electron microscopic study. J Comp Neurol.262:563–577

Swanson LW, Petrovich GD (1998) What is the amygdala? Trends Neurosci 21:323–331

Tamamaki N, Fujimori KE, Takauji R (1997) Origin and route of tangentially migrating neurons in the developing neocortical intermediate zone. J Neurosci 17:8313–8323

Tashiro Y, Drake JM (1998) Reversibility of functionally endured neurotransmitter systems with shunt placement in hydrocephalic rats: implications for intellectual impairment in hydrocephalus. J Neurosurg 88:709–717

Tashiro Y, Chakrabortty S, Drake JM, Hattori T (1997) Progressive loss of glutamic acid decarboxylase, parvalbumin, and calbindin D28K immunoreactive neurons in the cerebral cortex and hippocampus of adult rat with experimental hydrocephalus. J Neurosurg 86:263–271

Ten Donkelaar HJ (1988) Evolution of the red nucleus and rubospinal tract. Behav Brain Res 28:9–20

Ulfig N (1989) Configuration of the magnocellular nuclei in the basal forebrain of the human adult. Acta Anat 134:100–105

Ulfig N (1999) Introduction to histochemical studies on human fetal brain development. Microsc Res Tech 45:339–340

Ulfig N (2000a) The ganglionic eminence – new vistas. Trends Neurosci 23:530

Ulfig N (2000b) Transiente Charakteristika des fetalen Gehirns und ihre Bedeutung für ZNS-Komplikationen des Frühgeborenen. In: Friese K, Plath C, Briese V (eds) Frühgeburt und Frühgeborenes. Eine interdisziplinäre Aufgabe. Springer, Berlin Heidelberg New York, pp 3–17

Ulfig N (2001) Expression of calbindin and calretinin in the human ganglionic eminence. Pediatr Neurol 24:357–360

Ulfig N (2002) Neuronal vacuolation in the basal nucleus of Meynert caused by fetal hydrocephalus Pediatr Neurosurg (in press)

Ulfig N, Chan WY (2001) Differential expression of calcium-binding proteins in the red nucleus of the developing and adult human brain. Anat Embryol 203:95–108

Ulfig N, Chan WY (2002a) Axonal patterns in the prosencephalon of the human developing brain. Neuroembryology 1:4–16

Ulfig N, Chan WY (2002b) Expression of AKAP79 and synaptophysin in the developing human red nucleus. Neurosignals (in press)

Ulfig N, Chan WY (2002b) Axonal patterns in the prosencephalon of the human developing brain. Neuroembryology 1:4–16

Ulfig N, Rupp M, Braak E, Braak A (1989) Hypothalamic tuberomamillary nucleus in man: nuclear configuration, cell types and parvalbumin-immunoreactive neurons. Eur J Neurosci Suppl 2:216

Ulfig N, Nickel J, Bohl J (1998a) Monoclonal antibodies SMI 311 and SMI 312 as tools to investigate the maturation of nerve cells and axonal patterns in human fetal brain. Cell Tissue Res 291:433–443

Ulfig N, Nickel J, Bohl J (1998b) Transient features of the thalamic reticular nucleus in the human foetal brain. Eur J Neurosci 10:3773–3784

Ulfig N, Setzer M, Bohl J (1998c) Transient architectonic features in the basolateral amygdala of the human fetal brain. Acta Anat 163:99–112

Ulfig N, Tietz B, Bohl J (1999) Alterations in the organization of the isocortical layer I in trisomy 22. Neurosci Res 33:119–125

Ulfig N, Setzer M, Neudörfer F, Bohl J (2000d) Distribution of SNAP-25 in transient neuronal circuitries of the developing human forebrain. Neuroreport 11:1259–1263

Ulfig N, Feldhaus C, Bohl J (2000a) Transient expression of synaptogyrin in the ganglionic eminence of the human fetal brain. Ann Anat 182:505–508

Ulfig N, Feldhaus C, Setzer M, Bohl J (2000b) Expression of MAP1a and MAP1b in the ganglionic eminence and the internal capsule of the human fetal brain. Neurosci Res 38:397–405

Ulfig N, Neudörfer F, Bohl J (2000c) Transient structures of the human fetal brain: subplate, thalamic reticular complex, ganglionic eminence. Histol Histopathol 15:771–790

Ulfig N, Setzer M, Neudörfer F, Bohl J (2000d) Distribution of SNAP-25 in transient neuronal circuitries of the developing human forebrain. Neuroreport 11:1259–1263

Ulfig N, Setzer M, Neudörfer F, Saretzki U (2000e) Changing distribution patterns of synaptophysin-immunoreactive structures in the human dorsal striatum of the fetal brain. Anat Rec 258:198–209

Ulfig N, Neudörfer F, Bohl J (2001a) Development-related expression of AKAP79 in the striatal compartments of the human brain. Cells Tissues Organs 168:319–329

Ulfig N, Szabo A, Bohl J (2001b) Effect of fetal hydrocephalus on the distribution patterns of calcium-binding proteins in the human occipital cortex. Pediatr Neurosurg 34:20–32

Van der Kooy D, Fishell G, Krushel LA, Johnston JG (1987) The development of striatal compartments: from proliferation to patches. In: Carpenter MB, Jayaraman A (eds) The basal ganglia. structures and concepts-current concepts. Plenum Press, New York, pp 81–98

Verney C (1999) Distribution of the catecholaminergic neurons in the central nervous system of the human embryos and fetuses. Microsc Res Tech 46:24–47

Verney C, Febvret-Muzerelle A, Gaspar P (1995) Early postnatal changes of the dopaminergic mesencephalic neurons in the weaver mutant mouse. Brain Res Dev Brain Res 89:115–119

Volpe JJ (1995) Neurology of the newborn. Saunders, Philadelphia

Volpe JJ (1996) Subplate neurons – missing link in brain of the premature infant? Pediatrics 97:112–113

Von Monakow C (1909) Der rote Kern, die Haube und die Regio hypothalamica bei einigen Säugetieren und beim Menschen. Vergleichende anatomische, normal-anatomische, experimentell und pathologisch-anatomische Untersuchungen. I. Teil: Anatomisches und Experimentelles Arb Hirnanat Inst Zürich 3:49–267

Voogd J, Nieuwenhuys R, van Dongen PAM, ten Donkelaar HJ (1998) Mammals. In: Nieuwenhuys R, ten Donkelaar HJ, Nicholson C (eds) The central nervous system of vertebrates, vol. 3. Springer, Berlin Heidelberg New York, pp 1637–2097

Voorn P, Kalsbeek A, Jorritsma-Byham B, Groenewegen HJ (1988) The pre- and postnatal development of the dopaminergic cell groups in the ventral mesencephalon and the dopaminergic innervation of the striatum of the rat. Neuroscience 25:857–887

Walicke PA, Patterson PH (1981) On the role of Ca2+ in the transmitter choice made by cultured sympathetic neurons. J Neurosci 1:343–350

Wang YJ, Liu CL, Tseng GF (1996) Compartmentalization of calbindin and parvalbumin in different parts of rat rubrospinal neurons. Neuroscience 74:427–434

Winkler C (1929) Opera omnia. Tome IX. De Erven F. Bohn, Haarlem

Wood JG, Martin S, Price DJ (1992) Evidence that the earliest generated cells of the murine cerebral cortex form a transient population in the subplate and marginal zone. Brain Res Dev Brain Res 66:137–140

Wright LC, McAllister JP 2nd, Katz SD, Miller DW, Lovely TJ, Salotto AG, Wolfson BJ (1990–1991) Cytological and cytoarchitectural changes in the feline cerebral cortex during experimental infantile hydrocephalus. Pediatr Neurosurg 16:139–155

Zecevic N (1993) Cellular composition of the telencephalic wall in human embryos. Early Human Dev 32:131–149

Subject Index